NEWS CORPORATION, TECHNOLOGY AND THE WORKPLACE
Global Strategies, Local Change

This book, which includes extensive interview material and primary research, examines technological innovation and workplace restructuring carried out by News Corporation in its newspaper holdings in Britain, the United States and Australia. Timothy Marjoribanks finds that while some outcomes at various local sites were similar, many were dramatically different. His study reveals that the nature of existing social relations in a particular location has a major impact on workplace reforms. The book finds that the prevailing balance of power between trade unions and workers, management and employers, and the role of the state in these relationships are the most influential factors in determining the course of events. Significantly, it emphasises the importance of analysing the connections between events occurring locally, nationally and globally if we are to understand the growing influence of corporate actors such as News Corporation.

Timothy Marjoribanks is T. R. Ashworth Lecturer in Sociology at the University of Melbourne. A graduate of Harvard University with a PhD in sociology, his current research examines corporate and workplace governance in the media and entertainment industries.

D1358152

For Karen and Rebecca

NEWS CORPORATION, TECHNOLOGY AND THE WORKPLACE

Global Strategies, Local Change

TIMOTHY MARJORIBANKS

University of Melbourne

CAMBRIDGE
UNIVERSITY PRESS

PUBLISHED BY THE PRESS SYNDICATE OF THE UNIVERSITY OF CAMBRIDGE
The Pitt Building, Trumpington Street, Cambridge, United Kingdom

CAMBRIDGE UNIVERSITY PRESS
The Edinburgh Building, Cambridge CB2 2RU, UK   http://www.cup.cam.ac.uk
40 West 20th Street, New York, NY 10011–4211, USA http://www.cup.org
10 Stamford Road, Oakleigh, 3166, Australia   http://www.cup.edu.au
Ruiz de Alarcón 13, 28014, Madrid, Spain

First published 2000

Printed in Singapore by Green Giant Press Pte Ltd

Typeface Baskerville (Adobe) 10/12 pt.   *System* Penta [MT]

A catalogue record for this book is available from the British Library

National Library of Australia Cataloguing in Publication data
Marjoribanks, Timothy Kevin.
News Corporation, technology and the workplace : global
strategies, local change.
Bibliography.
Includes index.
ISBN 0 521 77280 X.
ISBN 0 521 77535 3 (pbk).
1. News Corporation – Management. 2. Press – Effect of
technological innovation on. 3. Newspaper employees –
Effect of technological innovations on. 4. Organizational
change. 5. Trade-unions – Newspaper employees. 6.
Industrial management. 7. Technological innovations –
Social aspects. I. Title.
338.064

ISBN 0 521 77280 X hardback
ISBN 0 521 77535 3 paperback

Contents

Tables

Figures

Acknowledgements

In conducting field research for this book, including interviews and site visits, I was continually and wonderfully surprised as to how willingly people gave of their time. The book would have been substantially different had it not been for the generosity of members of the newspaper industry in countries and workplaces around the world. Their voices and experiences are central to the story told, and I am grateful to them. In Britain, I owe special thanks to David Bell of the *Financial Times* (and to David Burnham in the United States), and to members of the GPMU located in Bedford. David Bell was extremely open to my requests for information, and through his efforts I was introduced to a wide cross-section of newspaper industry people. Members of the GPMU provided me with valuable resources, as well as personal accounts of their experiences of the influence of technology on work. Motoko Rich, a journalist at the *Financial Times* when this research was conducted, was another who introduced me to important individuals in the industry. To those I have named, and to the many I have not named but who allowed themselves to be interviewed, including, in particular, members of the *Advertiser*, the PKIU, and the NUJ, I thank you.

When this book was in its dissertation stage at Harvard University, a number of individuals provided invaluable guidance in the discipline of sociology. Theda Skocpol, the head of my dissertation committee, was a thoughtful and critical advisor, while also being a constantly supportive and reassuring voice through the long process of research and writing. John Campbell introduced me to important theoretical debates around the social organisation of the economy, many of which appear here. John was also a close and critical reader of my work, providing invaluable advice on both the structure and content. Mary-Jo DelVecchio Good was to become not only a highly valued advisor, but also a very close friend. Her role and influence as a teacher, critical reader and friend have provided me with the sort of professional mentorship and personal camaraderie that develops only very rarely. I also thank Juliet Schor, who provided important criticisms of my work in the early stages of this project, and whose influence is still evident in these pages. Kevin Marjoribanks and Karen Farquharson both read many drafts of this work, and were critical yet friendly readers. Jim Ellis, John Glenn, David Kang, Jeff Marinacci and Francesco Duina also provided valuable input at various stages of the research process.

Colleagues in the Sociology Program and the Department of Political Science at The University of Melbourne have also been a source of encouragement and support. I owe particular thanks to Ann Capling for taking a punt and introducing the work of a new colleague to Cambridge.

Phillipa McGuinness and the staff at Cambridge University Press have been wonderfully supportive of this project, providing assistance, advice and enthusiastic support throughout. Thank you also to the two anonymous readers whose critical but sympathetic comments helped me greatly.

Throughout my life, my family has been my greatest source of support, guidance, friendship and love. My father Kevin, my mother Janice and my sister Genevieve have always been there for me, in times of celebration and happiness, but also at times of doubt and uncertainty. Their strength of character and their loving advice, related to both academic and personal matters, have been at the foundation of this work. I have no hesitation in saying that without them this book would never have been written. As my family has been such a source of love and friendship, so my marriage to Karen in 1996 was to be a pivotal moment in my life. Karen's advice and sociological insights were vital to the development of this research, and her loving support and input have been crucial. Our daughter Rebecca was born as I was writing up the final version of the book, and she has brought with her wonderful joy. To my family, I owe my greatest thanks of all.

Research for this book in its dissertation stage was assisted by a Frank Knox Memorial Fellowship, awarded over two years by the President and Fellows of Harvard College, in consultation with the Australian Vice-Chancellors' Committee, and by a Harvard University Graduate Society Dissertation Completion Fellowship.

Selected extracts from *News and Fair Facts, The Australian Print Media Industry* (Australian House of Representatives Select Committee Report on the Print Media, Australian Government Publishing Service, 1992) and from *Termination, Change and Redundancy Case* (Commonwealth Arbitration Reports, Australian Government Publishing Service, 1984). Commonwealth of Australia copyright, reproduced by permission.

Abbreviations

ACAS	Advisory, Conciliation and Arbitration Service
ACTU	Australian Council of Trade Unions
AFL–CIO	American Federation of Labor–Congress of Industrial Organizations
AFMEU	Automotive, Food, Metals, and Engineering Union
AHRSC	Australian House of Representatives Select Committee
AJA	Australian Journalists' Association
ALP	Australian Labor Party
AMWU	Australian Manufacturing Workers' Union
ANPA	American Newspaper Publishers' Association
BCA	Business Council of Australia
CAI	Confederation of Australian Industry
CBI	Confederation of British Industry
CWA	Communication Workers of America
EETPU	Electrical, Electronic, Telecommunications and Plumbers' Union
ENS	Electronic Newspaper System
ETU	Electrical Trades Union
FCU	Federated Clerks' Union
FT	*Financial Times*
GPMU	Graphical, Paper and Media Union
Guild	The Newspaper Guild
ITU	International Typographical Union
JSC	Joint Standing Committee for National Newspapers
NGA	National Graphical Association
NLRA	National Labor Relations Act
NLRB	National Labor Relations Board
NPA	Newspaper Proprietors' Association
NPA	Newspaper Publishers' Association
NTA	New Technology Agreement
NUJ	National Union of Journalists
PANPA	Pacific Area Newspaper Publishers' Association
PATCO	Professional Air Traffic Controllers' Organization

PATEFA	Printing and Allied Trades Employers' Federation of Australia
PKIU	Printing and Kindred Industries Union
PTJ	Printing Trades Journal
PVT	Personal Video Terminal
RSI	Repetitive Strain Injury
SCP	St Clements' Press
SOGAT	Society of Graphical and Allied Trades
TCR	Termination, Change and Redundancy
TDC	Trade Development Council
TGWU	Transport and General Workers' Union
TNT	Thomas National Transport
TUC	Trades Union Congress
VDT	Video Display Terminal

INTRODUCTION

Tales from the Workplace

On a grey and overcast Saturday in late January 1995 a crowd of some 500 people assembled at Tower Hill underground station near the River Thames in London. The people in the crowd milled about, talking, greeting friends, reading pamphlets and trying to keep warm, while some distributed and held banners representing labour organisations including the Graphical, Paper and Media Union (GPMU), and the National Union of Journalists (NUJ), while others represented political parties. Suddenly, those with banners were called to the front by yellow and orange bibbed marshals, announcements were shouted, and people began to march. The march proceeded along the street with police – some on horseback – ensuring that no one spilled over onto the median strip. After an initial period of quiet, chants filled the air. People shouted and responded 'Rupert Murdoch, hear us shout! You can't keep the unions out!', 'What do we want? Union rights! When do we want them? Now!', 'Major! Major! Major! Out! Out! Out!'. The march wound along the streets of London until the destination was in sight – the buildings of the media organisation News International at Wapping in the Docklands area of London. As the people passed these buildings the chants grew louder, and the passion and anger of those involved in the march was evident. Numerous police and security guards protected the entrance to the Wapping site to ward off any attempts to enter the heavily fenced premises. The people marched past the building, staring at it and at the people inside, and headed towards a small park across the street from the buildings owned by News International. At the park some people huddled under umbrellas in the falling drizzle and others collected money, while trade unionists, workers in the newspaper industry and politicians gave short speeches to the assembled gathering. One worker recounted her story of how

1

she had been sacked recently by the company for trying to organise with co-workers a request for more ergonomically sound equipment. Another speaker, the General Secretary of the NUJ, argued that although not officially recognised within the plant at News International, unions were still a presence, and would not surrender to the attempts of the company to remove all trace of them. The cheers were loud, and there was a sense of defiant optimism.

To the casual observer, this march may have appeared as simply another labour demonstration. Placed within its historical context, however, the march assumes much greater significance. Nine years previously, on 26 January 1986, 5,500 employees of News International had been dismissed as the company undertook a process of workplace restructuring that was to result in a dispute which lasted for more than twelve months. News International relocated from the traditional site of national newspaper production in Britain in Fleet Street to a new site at Wapping, in a move that was soon to be followed by other major British newspapers.

That dispute in 1986, its origins and its outcomes, which are the focal concerns of this study, subsequently altered not only workplace and industrial relations in the newspaper industry, but also influenced the organisation of workplace relations throughout British industry. It was a dispute ostensibly involving only the introduction of technology into the newspaper facilities of News International, but it raised fundamental questions about workplace power, such as: who is to control technological innovation in the workplace? How is technology to be introduced? Who is to benefit from the introduction of technology? How are events in the workplace influenced by institutions and relations in the broader society? It is these broader questions about the relations between technological innovation and workplace reorganisation that I examine in this study.

News International is, however, just one part of the global media corporation, News Corporation. It soon became apparent that the effects of Wapping were to be experienced not only in Britain, but in other countries where News Corporation had interests, most notably in Australia and the United States. In the mid to late 1980s, News Limited in Australia began an expansive program of technological innovation across all of its newspapers in cities including Adelaide, Sydney, Melbourne and Brisbane. It was in Adelaide that Rupert Murdoch, the Chief Executive of News Corporation, began his media career with the now defunct Adelaide *News*. The program of innovation in Australia involved the introduction of computer-based technologies and also relocation of production work as had occurred in London. Throughout this process of technological innovation and workplace reorganisation, the shadow of Wapping was never far away. Workers and management at the News Corporation-owned Adelaide *Advertiser*, for example, talked of their new production facility as 'Wapping South'. At the same time, technologically-related workplace reorganisation was carried out at the *Advertiser* without the mass sackings or the twelve-month dispute of Wapping. A further question that this study addresses is why was the relationship between technological innovation and workplace reorganisation different at the *Advertiser* to that at the Wapping site of News International?

News America Publishing, the United States subsidiary of News Corporation, was also to be affected by Wapping. In his Walter Wriston lecture in New York in 1989, Rupert Murdoch proclaimed that with the establishment of a non-union workplace at Wapping, the New York newspaper unions were now the most difficult in the world (Murdoch 1989a). Throughout the 1980s and into the 1990s, newspapers owned by News America including the *New York Post* and the *Boston Herald* were to be involved in conflicts and disputes over the relative workplace influence of management and unions, often related to technological questions. As part of the comparative analysis of this study, I will explore to what extent the relationship between new technology and workplace reorganisation was different in the US newspaper holdings of News Corporation to that of Wapping.

While a number of theoretical positions have been developed to examine the relations between the introduction of new technology and workplace reorganisation, I will build on these previous frameworks to generate a model which I have labelled the 'relational model'. The model states that control over the introduction of new technology into the workplace, and the influence that this has on workplace relations, is dependent on the prevailing balance of power between trade unions and workers, management and employers, and the state. Further, this balance of power can be understood only by considering the institutional, political, social and economic contexts in which the actors are situated. It is this model that I use to examine the complexity of factors that have influenced the relationship between the introduction of technology and workplace reorganisation at the newspaper sites of News Corporation in Britain, Australia and the United States.

PART I

Technological Innovation and Workplace Reorganisation: The Newspaper Industry

CHAPTER 1

Global Technology and the Local Workplace: A Theoretical Debate

The influence of technology on all aspects of society – on the people, the institutions, and relations within societies – is being both celebrated and criticised in the late twentieth century. Technological developments in computers, telecommunications, satellites and fibre optics are proclaimed by many as heralding a new, more productive and economically efficient society. Two areas where the influence of computer technology is particularly prevalent are in the organisation of work and the workplace, and the global expansion of the media and communications industries. Walter Wriston, the former Chair and Chief Executive Officer of the US-based Citibank, has been reported as claiming:

> [T]he revolution in information technology is threatening the sovereignty of nation-states. The global economy is an information economy in which information is the most important factor of production of goods and services. Intellectual capital (knowledge and information) is the new source of wealth for nations (not manufactured goods or agricultural products, but information) (Stephens 1995:20).

Commenting on the interaction between technology, the media and world events, Rupert Murdoch, the Chair and Chief Executive of the global media company News Corporation Limited, observed that over a period of thirty years he had witnessed the media playing a fundamental role in liberating nations, and in transforming politics and culture around the world (Murdoch 1993:8).

While an apparent internationalisation of media linked to technological developments has occurred, a globalisation of the world economy has also taken place which is linked fundamentally to the dispersion of computer and other electronic and digital technology

(Freeman, Soete and Efendioglu 1995; Lash and Urry 1994). Globalisation is defined by the Organisation for Economic Cooperation and Development (OECD) as:

> an evolving pattern of cross-border activities of firms, involving international investment, trade and collaboration for the purposes of product development, production and sourcing, and marketing. It is driven by firm strategies to exploit competitive advantages internationally, use favourable local inputs and infrastructure, and locate in final markets (OECD 1994:28).

Among the factors which characterise the globalisation of economic relations are the further development of transnational corporations; the internationalisation of technologies; increased trade between nations; and financial flows linked to the development of global banking and the liberalisation of trade, investment, and capital movements. According to Fagan and Webber (1994), however, the social impacts of globalisation have been sharply uneven at the local level. They argue that there is a poor understanding of the interactions between the global and the local, with the impact of economic restructuring being experienced differentially at global, national and local levels (for further discussion of globalisation see Berger and Dore 1996; Boyer and Drache 1996; Capling, Considine and Crozier 1998; Hollingsworth and Boyer 1997; Lash and Urry 1994; Sklair 1995). In this regard, the local workplace is a crucial area of investigation when the impact of globalisation on workplace reorganisation is being considered.

The connections between economic globalisation, global technology and local workplace practice raise significant matters for analysis. Computer technologies do make it possible for some sections of the 'global population' to have access to more information than ever before, and they play an important role in enabling various companies to expand globally. Such technology, however, is experienced unevenly at the workplace where it has a fundamental and direct impact on the lives of people. While developments in information and communications technology are related to gains in high technology and science-based manufacturing employment in OECD countries, evidence suggests that those gains are offset by a decline in employment in other manufacturing industries associated also with technological developments. According to Freeman, Soete and Efendioglu:

> [T]he rapid diffusion of ICT [Information and Communications Technology] has led – and continues to lead – to a substantial 'exclusion' of

large parts of the labour force, either unskilled or wrongly skilled and inca-
pable of retraining. This bias in the demand for labour, which has only
emerged over the past 10 to 15 years, is likely to become much more
pronounced in the rest of the 1990s (1995:600).

Such biases affect not only blue collar workers in traditional manufacturing industries, but also white collar workers with skills no
longer deemed appropriate or essential.

These contradictory effects of information and communication
technologies are also experienced in specific industries, and they
have had a dramatic impact on the media industry. As new technologies create opportunities for global information expansion, for the
development of new global media 'empires', and for new skill requirements, they also make redundant whole classes of activities,
including craft and manual, and even professional, work. The consequences of technology for those individuals made redundant by
technological development present another side to the celebration of
the new information-based economy by business leaders such as
Wriston and Murdoch. Such developments and outcomes raise questions about the negotiation processes related to technological
innovation and subsequent workplace reorganisation in the media
industry. In particular, questions arise as to who controls the process
of innovation and reorganisation, how this control is exercised, and
who benefits from the processes.

In this study, the social and technological developments in the
global, national and local arenas will be examined by exploring the
contradictory influences of technology in the media industry. These
contradictions will be investigated by examining relationships
between developments in the global dispersion of computer technology and the reorganisation of workplace practices and relations in
the newspaper industry in Australia, Britain and the United States
from the 1970s to the 1990s. In particular, relationships are examined between technological innovation and workplace reorganisation
in newspapers owned by News Corporation. The late twentieth
century has been tumultuous for newspaper production as the craft
basis of the industry and the relations which had built up around
that craft basis have been challenged by the development and introduction of computer-based information technologies (Cahill 1993).
As well as leading to improvements in the physical presentation of
newspapers, through the use of advanced printing techniques incorporating colour, such technological advances have challenged
existing workplace practices and relations. In particular, workplace
demarcations in the pre-press stages of production have been

rendered potentially superfluous by technological developments which allow these pre-press processes to become integrated. The development of such technology has not resulted, however, in its immediate or universal adoption in newspaper workplaces. Factors including economic, political and institutional relations, workplace organisation, and the actions and perceptions of employers and employees, have interacted to influence the processes of technological innovation and adaptation in newspaper organisations.

The Social Organisation of Capitalist Economies

Any analysis of the relationship between the introduction of new technology and workplace reorganisation must consider transformations in the social organisation of the societies in which those workplaces are located. A commonly articulated view is that the post-World War Two period was one which witnessed the consolidation of Fordist systems of mass production based on moving assembly line techniques and operated by the semi-skilled labour of the mass worker. Taylorist principles of work organisation also predominated in this period, resulting in the standardisation of work practices and a separation between the conception and execution of work. This system of mass production also required mass consumption to remain viable, and as a result many social developments were introduced to support the creation of mass consumer markets. These social developments included Keynesian macroeconomic policies which acknowledged the role of the government in maintaining levels of domestic demand for mass-produced goods; the development of collective agreements over wages and workplace conditions between employers and labour with the support of the state; minimum wage legislation; and expanded welfare benefits. Such developments were associated with economies in which the incomes of wage-earners rose with productivity and high levels of consumption were achieved. The combination of mass production and mass consumption, which produced a general pattern of social organisation in capitalist societies, has become known as the Golden Age of Capitalism (Armstrong, Glyn and Harrison 1984; Capling, Considine and Crozier 1998; Glyn et al. 1990; Harvey 1990; Hollingsworth and Boyer 1997; Jessop 1991; Marglin 1990; Marglin and Schor 1990; Schor and You 1995).

In the late twentieth century, debate has emerged over whether social relations in capitalist economies are shifting from mass production and distribution techniques to more specialised and discrete modes of production. Much of the discussion concerns the precise nature of these shifts and their consequences for workplace organisation. A number of competing frameworks have been developed to examine the current period, often with a particular focus on workplace relations. These include the end of organised capitalism (Lash and Urry 1987), post-Fordism (Mathews 1989), the new competition (Best 1990), flexible specialisation (Piore and Sabel 1984), post-industrialism (Block 1990), the network society (Castells 1996), and the information economy (Poster 1995; Webster 1995). Such theoretical orientations suggest that contemporary technology and workplace developments offer a unique opportunity for workplace reorganisation in which workers on the shopfloor will be able to exert greater control over their work lives. Fred Block, a leading theorist of post-industrial society, discusses the possibilities of organisational learning associated with new technologies in which workplace reorganisation can be focused on a 'commitment to employment security and career development, democratisation of decision-making, and the establishment of employees as stakeholders' (Block 1990:203). In contrast, research by theorists including Hyman (1988; 1989), Milkman (1997), Pollert (1988; see also Gilbert, Burrows and Pollert 1992), and Wood (1989) suggests that while technologically-related workplace reorganisation is currently occurring, it does not represent a fundamental reform in the organisation of the workplace but rather constitutes the intensification of pre-existing class and workplace relations within those economies. They observe that rather than providing for greater worker participation in the workplace, new technologies are being used to increase the control of management over the workforce and workplace organisation.

In addition to the impact of new technologies, the social organisation of capitalist societies is being influenced by the structural reorganisation which is occurring in international and national economies. Processes of structural reorganisation include increasing integration of national economies on a global scale; shifts in national economies from manufacturing to service industries; increasing rates of product and technology innovation; and the reorganisation of corporate structures. These forms of economic restructuring have influenced the institutional regulation of social relations. For those involved in the workplace, developments in social regulation include shifts in industrial relations systems from centralised to enterprise and individual bargaining and a withdrawal of the role of the state; employers demanding workplace and employment flexibility; abandonment of full employment policies by governments; and trends towards extending the role of the market (Crouch and Streeck 1997; Ferner and Hyman 1992; Glyn et al. 1990; Golden and Pontusson

1992; Hollingsworth and Boyer 1997; Probert 1993; Schor and You 1995).

This economic and social restructuring is argued to be intimately associated with the development of new technologies, including microelectronics, new materials technology, biotechnology, new energy resources, and computer systems (Bamber and Lansbury 1989; Block 1990; Castells 1996; Poster 1995; Webster 1995). While new technologies have been developed continuously this century, many propose that computer technology is not just a further stage in technological development, but in fact marks a fundamental advance in the possibilities of technology. Mathews (1989, 1994), for example, talks of new technology, such as computer systems, as having changed the face of production and of production relations, while Piore and Sabel (1984) discuss the potential of new technological forms for the development of a new stage of capitalism. Theorists including Poster (1995) go even further to argue that developments in information technologies are contributing to the emergence of a qualitatively new society. From such perspectives, computer technology carries with it profound possibilities for restructuring both the process of work and workplace relations, and for altering the forces which influence workplace relations. The introduction of computer technology is likely to affect employment conditions, the division of labour within and between jobs, and the capacities and relations of the respective representatives of labour and capital, and their relationship with the state. Such technology also operates globally, bringing previously geographically separate workplaces and actors together (Sklair 1995).

In investigations which examine the influence of technology on workplace reorganisation, and the related question of whether capitalist economies are entering a new stage of social organisation, disagreement exists as to the precise contours of such social reorganisation, and to what extent technology interacts with other forces to affect reform. By examining the newspaper industry, which has been fundamentally influenced by computer technology, this study attempts to enhance contemporary understanding of the relationship between the introduction of new technology and workplace reorganisation, and in so doing, contribute to the debate about the social organisation of capitalist economies.

The Local Workplace in the Newspaper Industry: Technological Developments

The mode of workplace organisation in the production of a newspaper is related to the type of technology available and in use at a particular workplace. Newspapers produced after the invention of movable type in 1450 by Gutenberg were manual operations, involving typesetting and presses operated by human labour. In the early nineteenth century, print workers adapted the steam engine to use with Gutenberg's press, and the development of a steam driven rotary press by Friedrich Koenig in 1814 made mass circulation possible. A major development in newspaper production occurred in 1886 when the *New York Tribune* became the first newspaper to cast lines of type on the Linotype machine created by Ottmar Mergenthaler (Moghdam 1978). Linotype composition made use of a hot metal process which was to become the dominant means of newspaper production on a global basis for some ninety years, and for even longer in some instances, despite subsequent innovations in 'cold type' photocomposition. The use of 'cold type' from the 1950s introduced photographic processes into the printing process, so that words were produced on film and then transferred to a printing plate. The next revolution in the printing process came with the introduction of computerised visual display terminals, resulting ultimately in electronic pagination. Electronic pagination is a process whereby the pages of a newspaper are assembled electronically on a computer screen, with sub-editors assuming a central role in the design of page layout, and are then transmitted directly via computer to printing plates (Cockburn 1991; Conley 1997; Emery, Emery with Roberts 1996).

These recent innovations have had dramatic repercussions for workplace organisation and employment in the newspaper industry. Historically, the newspaper industry has been notable for its strict demarcations and specialisation of tasks between various crafts and non-crafts. The development of computer-based pagination, however, has rendered many of the craft skills potentially obsolete. As the Printing and Kindred Industries Union's (PKIU) *Australian Printing Industry Yearbook* indicated in relation to the printing industry in general:

> The digitalisation of information [which occurred in the 1970s and 1980s] has meant the development of equipment which dispenses with the need to separate the origination process into a number of discrete operations, and allows in its extreme form the preparation of graphic arts quality work by an author ... Moreover, the links possible with VDUs [visual display units], word processors and other elements of information technology allow a considerable degree of flexibility in the preparation and transmission of print-ready material. In effect, these front-end systems allow page make-up to take place at a much earlier stage than ever before. These pages can then be transmitted via telecommunications systems in a wide variety of ways to any location, whether national or international (Cahill 1993:21).

Such technological breakthroughs offer profound possibilities for reorganisation in the workplace. In particular, job demarcations and skill levels can be redefined. Reporting on the print media industry, an Australian House of Representatives Select Committee (AHRSC) commented:

> The main impact of these [technological] innovations appears to have been on the production and distribution processes and to a lesser extent on information gathering ... [V]arious traditional activities have been totally replaced by the use of computers. Phototypesetting, for example, has replaced the traditional linotype setters, proofreaders and compositors. Virtually all these functions are now performed by journalists and sub-editors on a computer terminal. Similarly, the transmission of finished pages by facsimile does away with the need to transport bulky finished newspaper copies (1992: 337).

Many of the tasks previously performed by members of the newspaper workforce, ranging from setting the newspaper page to distributing pages for printing, can now be undertaken by journalists working at a computer terminal. The House of Representatives Select Committee went on to comment on these developments in more detail in the same report:

> The video display unit has become the tool of the trade for journalists and editors. Journalists prepare their stories directly on a VDU and may move those stories electronically between themselves and sub-editors. All the necessary text manipulations and layout (editing, correcting, transposing rearranging lines of print, placing advertisements, etc.) are also performed directly on VDUs. When the story or page layout is complete it is stored electronically ready to be recalled on a VDU screen for phototypesetting. The complete text for a newspaper is thus prepared at a central location with inputs from journalists and editors who are not necessarily in the same location. Finally, the complete text is distributed electronically to one or more locations for printing (AHRSC 1992:337–8).

These developments make it apparent that many of the jobs held by craft workers are threatened by the shift to computerised production in the newspaper workplace. Compositors have historically been among the most powerful craft workers in the newspaper industry, based on their skills in setting and composing the stories of journalists into pages for print. The development of technology which makes pagination by journalists possible can lead to the elimination of the composing room altogether, however, by creating a work system in which journalists or editors electronically send completed pages directly to the printing plant. Direct entry systems allow journalists and advertising staff to keystroke into computers and prepare material for printing themselves.

As well as influencing or even eliminating the work undertaken by compositors and readers, technology of this type also alters the work of journalists. In addition to collecting information and writing stories, technological capacities mean that journalists can be required to develop layout skills and other tasks that were previously the specialised domain of craft and editorial workers (Conley 1997; Emery, Emery with Roberts 1996).

While these developments suggest that the newspaper industry has undergone major reorganisation in the last twenty to thirty years, care is needed in making generalisations about relationships between technological innovation and reorganisation in workplace practices across, or even within, national settings. Factors including economic conditions and institutional settings, and the balance of relations between employers, trade unions, workers and the state are likely to influence the manner in which technology is introduced into specific newspaper workplaces, and its subsequent impact on workplace relations. If our understanding of the relationship between technological innovation and workplace reorganisation is to be enhanced, a theoretical perspective is required that explores the complexity of the factors that might influence such reorganisation.

Exploring Technological Innovation: Alternative Theoretical Perspectives

The relationship between technological innovation and workplace reorganisation is one that has created much debate in the social sciences. Much of the analysis has examined whether workplace actors have some choice or influence over technologically-related workplace reorganisation, and to what extent such reorganisation processes are constrained, or even determined, by pre-existing structures, including economic relations and technology. Influential theoretical perspectives that have emerged from this debate include technological determinism, labour process theory, and an institutional social choice approach. In the remainder of this chapter these approaches will be reviewed, and an alternative relational model of the relationship between technology and workplace reorganisation is proposed. Throughout the discussion of the theoretical models a number of path diagrams will be used to examine the relationships proposed by each model. These diagrams are only an approximation of what are complex sets of relations, however, they illustrate the major relations suggested by the theoretical orientations.

Technological Determinism

An early approach adopted to investigate the relationship between technology and workplace organisation stresses the independent influence of technology on the workplace. From this perspective, which is labelled technological determinism, it is assumed that there is 'a straightforward link between technical possibilities and their implementation' (Williams with Thorpe 1992:41; see also Grint 1998). As firms seek to improve their efficiency and productivity, they introduce new technology which independently influences the organisation of the workplace. It is proposed that 'the outcomes of technological change are determined by the technical capabilities of a given technology' (Clark et al. 1988:9). This orientation continues to be influential in explaining patterns of workplace and economic reorganisation associated with technology. In his biography of Rupert Murdoch, for example, Shawcross writes that '[o]ur current revolution [that is, the change from an industrial to an information society] is driven by microelectronic technology. Economics has become secondary as a force for change. Politics finishes a poor third' (Shawcross 1992b:6). As technology modifies the tasks to be carried out by a particular business organisation, it is proposed that the form of the organisation is also modified. Technology is considered to be an independent variable which has a decisive impact on organisational form. That is, 'technology tends to determine organisational arrangements and a particular pattern of management, structure of work, and industrial relations' (Lansbury and Bamber 1989:5).

Advocates of the technological determinism model indicate that the patterns of reorganisation that occur in the structure of management, in the organisation of work, and in industrial relations in the newspaper industry are determined by the type of technology that is introduced. Similar patterns would be expected to occur in the organisation of different firms with the introduction of similar technology. The relationships proposed by the model of technological determinism are summarised in Figure 1.1, which indicates that new technology has a direct effect on workplace organisation, and an indirect effect through patterns of management and management–labour relations.

While this approach has the merit of drawing attention to the need to analyse the specifics of technology when considering processes of workplace reorganisation, the theoretical perspective has a number of limitations. In particular, technological determinism does not accord sufficient attention to social and political relations in capitalist economies, and does not consider adequately the power relations that exist in industries and institutions which may influence relationships

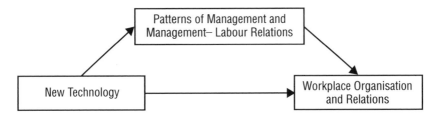

Figure 1.1 Technological Determinism Model

between technology and workplace reform. It is not possible in this model to explain variations in processes and outcomes of workplace reorganisation associated with similar technological innovations across a number of workplaces. Some of these limitations are addressed by labour process theory.

Labour Process Theory

An alternative approach used in the analysis of technological innovation and its impact on workplace organisation is labour process theory. Coming to prominence in the early 1970s with the publication of Harry Braverman's *Labor and Monopoly Capital* (1974), and grounded in the economic and political writings of Marx and Gramsci, labour process theory initially emphasised the de-skilling impact of new technology on work, and argued that the central strategy of management was to control the workforce. More specifically, the concept of de-skilling has been used in this approach to argue that 'management has consciously set out to reduce the knowledge and initiative of the individual worker, centralise the planning and direction of production in its own hands, and impose a fragmented and tightly supervised distribution of tasks on the shopfloor' (Tolliday and Zeitlin 1991:7; see also Williams with Thorpe 1992). The introduction of new technology was considered by Braverman to be a central means of achieving de-skilling, so as to increase management control of the labour process.

A set of ethnographic studies of the labour process carried out by Michael Burawoy, a sociologist at the University of California, Berkeley, enhanced the development of labour process theory. (For further contributions to labour process theory see Frances 1993; Grint 1998; Knights and Willmott 1990.) Burawoy claimed that 'The labour process ... must be understood in terms of specific combinations of force and consent [in the workplace] that elicit cooperation in the pursuit of profit' (Burawoy 1979:30). Burawoy examined how the production process is maintained through various combinations of coercion and consent specific to the shopfloor, always recognising that these shopfloor relations emerge from societal or global forces (Burawoy 1985).

To show how cooperation on the shopfloor is achieved, Burawoy developed the concept of production regime, by which he means:

> the process of production has two moments. First, the organisation of work has political and ideological effects – that is, as men and women transform raw materials into useful things, they also reproduce particular social relations as well as an experience of those relations. Second, alongside the organisation of work – that is, the labour process – there are distinctive political and ideological apparatuses of production which regulate production relations. The notion of production regime or, more specifically, factory regime embraces both these dimensions of production politics (1985:7–8).

Burawoy (1985) proposed that to understand production there is a need to examine the political and ideological, as well as the economic moment of production. That is, production regimes are 'combinations of persuasion and force which constitute the overall form of production under changing historical conditions' (Carroll 1990:395).

Using the concept of production regime, Burawoy develops a typology of factory forms based both on the combination of force and consent used to control relations within the factory, and on the link between external and internal regulation of workplace relations. He argues that the dominant form of factory regime into the early 1980s was the hegemonic form, in which state and factory apparatuses are institutionally separated but in which the state shapes the factory apparatus by stipulating, for example, mechanisms for the conduct and resolution of struggle (Burawoy 1985). Consent prevails within the specific factory in a hegemonic production regime, but coercion continues to coexist with such consent. To understand how shifts in the form of production regime occur, we need to examine the ways in which management and workers seek to maintain consensual relations on the shopfloor as broader economic, political, and social conditions develop. According to Burawoy, a shift in production regimes is occurring in the late twentieth century with the emergence of a regime of hegemonic despotism in which:

> the interests of capital and labour continue to be concretely coordinated, but where labour used to be granted concessions on the basis of the expansion of profits, it now makes concessions on the basis of the relative profitability of one capitalist vis-a-vis another ... The new despotism is the 'rational' tyranny of capital mobility over the collective worker. The reproduction of labour power is bound anew to the production process, but, rather than via the individual, the binding occurs at the level of the firm, region or even

nation-state. The fear of being fired is replaced by the fear of plant closure, transfer of operations, and plant disinvestment (1985:150).

It follows from Burawoy's analysis that in order to explain the dynamics of workplace reorganisation we need to examine the balance of force and consent in the shopfloor relations of workers among themselves, and in their relations with management. We need to be aware that workers will not always resist new technology which threatens their position in the workplace, but may endorse and sanction management plans concerning the introduction of technology (McLoughlin and Clark 1994). In addition, management will not necessarily force new technology onto the workforce, but may attempt to introduce new technology by gaining the consent of the workforce to the innovation process. Burawoy has developed a model of workplace organisation which recognises an active role for the different social actors, but which also seeks to locate those actions within the constraints of a particular stage of capitalist development.

In Figure 1.2, a model of the relationship between new technology and workplace reorganisation suggested by labour process theory is presented. In this model, new technology and social relations in an organisation are influenced by class relations, and the relationship between new technology and workplace reorganisation is mediated by the interaction that occurs between management and workers.

From labour process theory, the following propositions are generated regarding the relationship between the introduction of new technology and workplace organisation and relations:

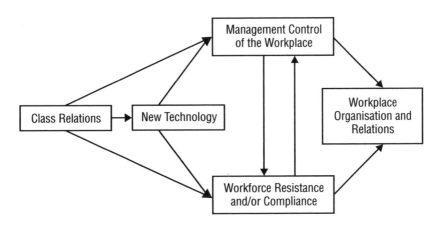

Figure 1.2 Labour Process Theory Model

1 The process of technological innovation and subsequent workplace restructuring is shaped primarily by the location of the particular workplace within a specific set of class relations.

2 Although management ultimately controls the process of introducing technology and reorganising the workplace in a capitalist economy, the workforce and its representatives play an active role in the process by either resisting or accommodating technological innovation on the shopfloor.

3 At a particular moment in the development of class relations, a particular set of workplace relations tends to predominate across industry.

4 The relationship between the introduction of technology and subsequent workplace reorganisation can in general terms be predicted from one workplace to the next, based on the stage of development of capitalism.

In a critical analysis of Burawoy's labour process model, Sturdy, Knights, and Willmott (1992) argued that more attention needs to be focused on resistance within the productive process. They also proposed that greater emphasis should be directed at the way in which social institutions – including those of the state – condition the organisation of the labour process. While the potential for resistance is possible within Burawoy's model, particularly with his use of the concept of hegemony as developed by Gramsci (1971), at times it disappears because of his concern to show the way in which consent is organised on the shopfloor. As a result, the possibility of accounting for the influence of technology on variations in the process and outcome of workplace reorganisation is restricted. Some of these limitations of the labour process model are addressed by the institutional social choice model.

Institutional Social Choice Model

An alternative theoretical perspective is the institutional social choice approach proposed by writers such as Tolliday and Zeitlin (Tolliday and Zeitlin 1991; Zeitlin 1985; Sabel and Zeitlin 1997; see also Immergut 1998; Kochan, Katz and McKersie 1986, 1994; Locke and Thelen 1995; Pierson 1994; Sabel 1982; Steinmo, Thelen and Long-streth 1992; Thomas 1994). While Burawoy works within a Marxist framework, and develops a model based on a periodisation of capitalism as a mode of social organisation, Tolliday and Zeitlin emphasise contingent circumstances and the role of institutions in shaping workplace relations. In particular, they reject the 'teleological model of capitalist development' which they see as inherent in the Marxist approach (Tolliday and Zeitlin 1991:9). When they consider the role of management, for example, they take issue with existing approaches which seem to allow for a degree of autonomy on the part of managerial actors, yet ultimately 'remain unwilling to concede a genuine scope for managerial choice in the face of the objective constraints imposed by the external environment' (Tolliday and Zeitlin 1991:12). Drawing on writers such as Weber, Geertz and Bourdieu for support, Tolliday and Zeitlin propose a more indeterminate approach in which 'interests should be regarded as inherently ambiguous, context-dependent and potentially incoherent' (1991:21). Through a series of case studies, they conclude that 'interests emerge from the interaction between social actors' prior interpretative framework and the specific situation in which they find themselves, a context which includes the discourses and practices of institutions such as employers' associations and trade unions as well as social and economic relationships' (Tolliday and Zeitlin 1991:21–2). That is, Tolliday and Zeitlin attempt to move beyond any conception of determinism which they suggest is inherent in other approaches, in particular Marxist approaches.

An important element of the institutional social choice approach is the emphasis on the role of the state. Again taking issue with Marxist orientations, which it is claimed do not sufficiently examine the role of the state, Tolliday and Zeitlin argue that there is a need for a direct theoretical focus on the state as a separate entity with a distinctive logic and a particular individual history (1991; Zeitlin 1985; see also Campbell and Lindberg 1990). In presenting this argument, Tolliday and Zeitlin draw specifically on Skocpol who has suggested that existing analyses of political, social and economic relations and transformations do not afford 'sufficient weight to state and party organisations as independent determinants of political conflicts and outcomes' (Skocpol 1980:199). She emphasises that state policies are not created in response to the needs of the capitalist economic system, but rather 'are shaped and reshaped through the struggles of politicians among themselves' (1980:200). Following this approach, Zeitlin rejects the idea that states or state leaders necessarily act for any particular class or interest group. Rather, he suggests that we can 'treat state leaders as independent actors with their own goals and priorities, and states themselves as historically formed institutional complexes with widely different structures and capacities' (1985:29). Zeitlin suggests an important limitation to the autonomy of the state as there will be occasions when the state's top executive and legal personnel are faced with competing demands in fulfilling their roles as state officials responsive to state needs, and in being leaders of political organisations which

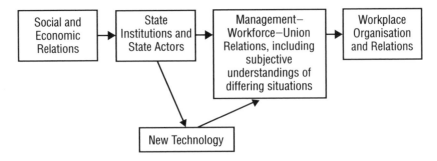

Figure 1.3 Institutional Social Choice Model

have to be aware of the interests of constituencies outside of the state apparatus (1985). That is, Zeitlin argues that there is an interaction between the state and society which influences the operation and capacity of the state.

By adopting a context-dependent approach, proponents of the institutional social choice model consider that they can explain apparently contradictory outcomes in the process of workplace reorganisation. They argue that processes of reorganisation need to be analysed as distinct instances in which the outcome will depend on which social interests are more influential at a particular time. The central argument of the institutional social choice model is that 'whatever the objective bases for conflict and cooperation between workers and employers, there can be no way to determine in advance which tendency will prove more fundamental' (Zeitlin 1985:8). Zeitlin claims that '[r]ecent studies of industrial conflict confirm that workers' behaviour depends not merely on their objective position in the division of labour, but equally on expectations they bring with them to the factory, on the behaviour of other groups, and on the wider historical and political context' (1985:8). In regard to technological innovation, Tolliday and Zeitlin indicate that 'it is the contingent outcome of historical events – themselves a product of the choices and expectations of the actors involved – rather than current or potential efficiency which shapes the trajectory of technological development and the resulting market structure' (1991:17).

Relationships suggested by the institutional social choice model are presented in Figure 1.3. In particular, state institutions and state actors mediate the associations between social and economic relations and new technology, management–workforce relations, and workplace reorganisation. State institutions and state actors also influence management–workforce relations directly and indirectly through the introduction of new technology. In addition, the relationship between the introduction of new technology and workplace reorganisation is mediated by management–workforce relations.

The institutional social choice approach suggests the following propositions:

1 In considering external processes which influence technological innovation and workplace reorganisation, the role of the state is fundamental. State policies and administrative decisions, legislation and judicial rulings, and the struggles between politicians and political parties all influence the context in which innovation and reorganisation occur.

2 The influence of social and economic relations on technological innovation and workplace reorganisation is mediated by these state institutions and political actors.

3 The manner in which new technology is introduced into the workplace, and the associated reorganisation of workplace relations, are the contingent outcome of historically developed workplace relations. The development of these workplace relations is based on interactions between management and workers, management and unions, and unions and workers.

4 While it is possible to identify in advance the actors who need to be analysed, including management, unions and workers, and the state, the behaviour and interests of these actors in specific situations cannot be predicted in advance. This includes the interests of actors who may appear to share a similar objective situation, but in fact behave differently in specific situations.

5 Analysis of the relation between technology and workplace reorganisation should proceed on a case by case basis, focusing on the specific circumstances of each workplace. There is no necessary correlation between events at one workplace and at another workplace, and generalisations across workplaces should not be made.

The institutional social choice approach directs attention to the role of the state in the process of innovation and reorganisation, and posits the importance of examining the historically developed capacity of the state as a constraint on what actions are possible for political and workplace actors. Nevertheless, at times this approach underplays the wider social relations and constraints within which the state and other actors are located. In particular, the economic conditions of capitalist societies, and the significance of classes and class organisations, are likely to be more constraining than is posited by Zeitlin and Tolliday (Immergut 1998; McEachern 1990). As a result, it is likely that in the institutional social choice model there is an overemphasis on the contingent elements of technological innovation and workplace reorganisation.

A Relational Model

In this analysis of the relationships between the introduction of new technology into the newspaper industry and workplace reorganisation a new model is presented in a bid to overcome some of the shortcomings of the previously discussed models. The relational model combines elements of labour process theory and the institutional social choice approach to explain workplace reorganisation. The theoretical orientations already discussed provide valuable insights into workplace relationships in the newspaper industry. As has been suggested, however, both labour process theory and the institutional social choice model possess certain limitations. In constructing an alternative framework to explain the relationship between technological innovation and workplace reorganisation a model was developed which recognises both the need for specific case studies while also acknowledging the constraints imposed by the location of those case studies within a particular social, economic and political configuration.

The relational model proposes that control over the introduction of new technology into the workplace, and the influence that this has on workplace relationships, is dependent on the prevailing balance of power between trade unions and workers, management and employers, and the state, and that this balance of power can be understood only by considering the political, social and economic contexts in which the actors are situated. That is, in comparison with previous models, state institutions and state actors have a direct impact on workplace reorganisation and economic, social and political relations directly influence the form of new technology. Such relationships are depicted in Figure 1.4.

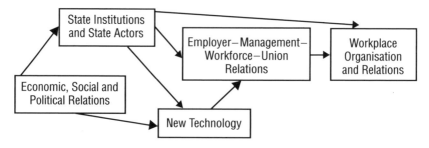

Figure 1.4 The Relational Model

The relational model which presents the general theoretical framework for this book suggests the following propositions for analysing relationships between the introduction of new technology and workplace organisation:

1 While employers and management control the relationship between the initial introduction of technology and workplace reorganisation through both their access to financial resources and the structural conditions of capitalism as an economic system, the workforce and its representatives influence that process through their capacity either to cooperate with or resist the introduction of technology in the workplace.
2 The institutions of the state, including the legislature, courts, and various departments or ministries and actors associated with the state, including political parties, influence the process of technological innovation through various processes intended either to facilitate or constrain the activities of employers, management, workers and unions. These processes may include legislation, legal decisions, and explicit or implicit policy.
3 Workplace reorganisation associated with the introduction of new technology occurs through interactions between these same actors and institutions, although the introduction of new technology into the workplace may influence the balance of power between those actors and institutions. That is, the introduction of technology and workplace reorganisation will be influenced by the historically developed and developing relations between management, labour and the state.
4 These historically developed relations are shaped and constrained by their location within a particular economic, political and social formation, and the processes and outcomes of technological development and workplace reorganisation are similarly shaped and constrained.

Conclusion

One of the central tasks of this book is to explore how the introduction of new technology has influenced workplace reorganisation in the newspaper industry at the level of the specific workplace. Another primary task is to examine to what extent the proposed relational model is an appropriate alternative theoretical model. In the analysis a number of case studies are presented which examine actual experiences of technological innovation and workplace reorganisation in newspapers in Britain, Australia, and the United States. The relational model proposes that investigations of the relationship between technological innovation and workplace reorganisation require an analysis of the institutional and societal settings in which the transformations are occurring. In the following chapter, I examine the institutional and societal settings of Britain, Australia and the United States from the 1970s to the 1990s, which provided the context for the dramatic transformations that were to occur in the newspaper industry.

CHAPTER 2

The Institutional and Societal Context: Britain, Australia and the United States

From the 1970s to the 1990s important developments occurred in the institutions and social, political and economic spheres in Britain and the United States that would enable employers to institute workplace reorganisation in the newspaper industry at the expense of union power. Both Britain and the United States have histories of strong union movements in the newspaper industry, however, by the end of the 1980s those unions had been severely weakened. At the same time, extensive reorganisation also occurred in the Australian newspaper industry, and yet unions continued to have a voice in negotiations over workplace reform. This chapter asks what occurred in the institutional and societal contexts of Britain and the United States for newspaper management to emerge into the 1990s in a position where they could operate almost independently of unions. And why did workplace reorganisation follow a different path in Australia?

The relational model, which provides the theoretical framework for this study, suggests that in order to understand the reorganisation of workplace relations associated with the introduction of new technology it is necessary to consider the institutions and the political, social and economic contexts in which the actors are situated. This chapter examines the specific contexts of Britain, Australia and the United States from the 1980s to the 1990s. It will show that although the three countries have much in common, their specific situations regarding workplace relations are quite different, and this had an important effect on how technological innovation occurred in the strongly unionised newspaper printing industry in each country.

The Context for the Restructuring of Industry: Britain

Economic conditions underwent a number of important shifts in Britain in the 1980s. Employment in the manufacturing sector fell from 30 per cent of total employment in 1980 to 23 per cent of total employment in 1990 (Edwards et al. 1992; Millward et al. 1992). In addition, and continuing a post-World War Two trend, Britain's share of world exports in manufactures declined from 21 per cent in 1953 to 10 per cent in 1990. The year 1983 was especially important, both economically and symbolically, as it marked the first time since the industrial revolution when more goods were imported into Britain than were exported (Edwards et al. 1992). At the same time, there was a growth of 12 per cent in employment in the financial services sector and in the services sector from 1984 to 1990. Unemployment remained a critical political and economic issue throughout the 1980s and into the 1990s (Millward et al. 1992).

The political party in office from 1979 until 1997 was the Conservative Party, and its approach to economic, political, and social relations was crucial in influencing the context for the restructuring of British industry. While the election of a Labour government in 1997, under the leadership of Tony Blair, has suggested the possibility of reform, they have had to deal with a much transformed social and institutional landscape. In particular, the years of Conservative Party government were to have an enormous impact on the role of unions in the British workplace, and on the institutional regulation of union–management relations.

The Conservative Party Government: 1979 to 1997

During the 1980s the Conservative Party government, under the leadership of Prime Minister Margaret Thatcher, instituted a number of policies that radically altered the ways in which workplace relations in Britain were managed. At the 1975 Conservative Party conference, Thatcher, who was to become prime minister in May 1979, introduced a vision of society founded on the neo-liberal philosophy that the individual must be able to operate without constraint in a free market economy. Rather than being dictated to by the state, and by organised labour, as had allegedly occurred in post-World War Two Britain, Thatcher argued that all people must be free to develop their abilities as they choose (Thatcher 1977). The implication was that state regulation of business activity and industrial relations should be rolled back to allow individuals to exercise their individual entrepreneurial abilities (Wedderburn 1989).

Developing and adopting this neo-liberal philosophy occurred in the context of the collapse of the Social Contract between the 1974 Labour Party government and the trade union movement, under the leadership of the Trades Union Congress (TUC). The Social Contract was an incomes policy which enabled unions to be involved in government policy formation, with unions agreeing to wage restraint in exchange for increased state spending in areas including 'welfare, pensions, schools, and benefits for those not in the labour market' (McEachern 1990:132). Union dissatisfaction with the government's proposed wage increases under the Social Contract in 1978 resulted in a series of strikes and the so-called 1978–79 'Winter of Discontent', which was to mark the demise of the Labour government (McEachern 1990; Pelling 1987). Throughout this period, the Conservative Party criticised the Labour government for its policies regarding the trade union movement. In particular, it accused the Labour Party of failing to deliver jobs, of allowing inflation to run rampant, and of promoting wastefulness and a lack of individual initiative on the part of the population in general and the workforce in particular (Armstrong, Glyn and Harrison 1984). Criticising the role of both the state and trade unions within society, the Conservative Party was able to campaign in the 1979 general election on the basis of a pro-market, anti-state philosophy in which the individual was purported to be paramount (Pelling 1987).

With the resulting victory of the Conservative Party, organised labour and sections of the state were soon to experience the influence of government policy based on neo-liberal prescriptions. These policies were based on the proposition that the most effective way to run the British economy would be to step away from government control and embrace free market principles, representing a dramatic shift from postwar policy in Britain. The major policies of the Conservative Party government, at least in the first years of office, included a restrictive monetary policy, cutbacks in welfare provision, taxation cuts, programs of privatisation and deregulation, and the weakening of trade unions (Armstrong, Glyn and Harrison 1984). Through such policies, the government sought to change the patterns of relations between the state and society, so as to restore the pre-eminence of the individual in society at the expense of the state bureaucracy (Graham 1997; McEachern 1990).

Although the success of the government in achieving its goals has been the subject of debate (Graham 1997; McEachern 1990; Pierson 1994), the government did succeed in reducing the political, economic and social role of trade unions. It fulfilled this goal in a number of ways, including legislation, support for business, and by removing trade unions from participation in government bodies. For example, the government cut funds to, and in some instances abolished, various tripartite organisations which brought together the government, trade unions and business confederations to discuss and formulate policy and which, to some extent, 'legitimised the union movement' (Purcell 1993: 8; see also Markey 1984). The hostility of the Conservative Party government to such organisations was a significant policy reversal, representing a concerted attempt to reduce the influence of trade unions.

A critical development in British industrial relations involving the government arose in 1984 with the conflict between the National Coal Board, which was a government body, and the National Union of Mineworkers (NUM). Eventually, the government succeeded in diminishing the role of the NUM in a protracted dispute over the profitability and viability of the coal industry in Britain that lasted many months. The government's main strategy was to close government-owned mines that were designated as economically unviable, in the face of concerted opposition from the NUM and the British labour movement. In addition, the law was used to fine the union for contempt of court over picketing, and to sequestrate the funds of various regional branches of the union. The miners' strike which ensued signalled the government's commitment to its market reforms, sending a signal to employers in the private sector that they would have government support in disputes with the unions. Furthermore, the success of the government in challenging the powerful NUM sent a message to the rest of the unionised workforce that no industry was to be spared from reorganisation.

Technological Development Policy

The Conservative Party government also succeeded in minimising the influence that unions were to have in the development and introduction of technology, in particular at the workplace. Typically, the policy approach of the government in regard to technical development was described by the Department of Trade and Industry as 'technology neutral', to describe its position of eschewing a role for government in technological development. Sharp and Walker describe this policy as complementing 'the guiding assumption of the 1980s ... that the market economy *must* be left to its own devices' (1994:397). They suggested that the Conservative Party, through the course of the 1980s, expressed little interest in technical change, and did not consider the use or production of technology to be a concern for the government. Instead, it was to be the concern of industry to innovate and expand in a market economy.

It needs to be emphasised, however, that while not being directly involved in technological development, the government did provide means of support for private business to invest in the products of technological development and innovation in the 1980s. In particular, the government sought to provide an economic environment which would stimulate market forces, to allow for the adoption of technological innovation in the workplace. For example, the government gave assistance to sectors of business through the creation of Urban Development Corporations (UDCs) and Enterprise Zones, as well as through the provision of various business loans for investment in technology (McIlroy 1991). UDCs were created by the Department of Trade and Industry and the Department of Investment, to fulfil the role of buying inner-city land, preparing industrial infrastructure, developing land for private business, and then selling the project to private business (Jessop et al. 1987). Enterprise Zones were established by the government in economically depressed areas for the development of business, and carried with them incentives for business, including exemptions from normal planning procedures and from various tax rates (Brown 1988; Sharp and Walker 1994).

While attempting to create an infrastructure which would support and assist investment, the government also attempted to constrain the activities and operations of trade unions. According to the Conservative Party government, trade unions were an important part of the economic and social problems that had befallen Britain. As a result, their influence needed to be reduced, if not altogether eliminated. It has been noted that the influence of various tripartite government bodies was reduced in this period. In addition, the Conservative government rejected any role for formal incomes policies, which had been central to the policies of the previous Labour government. While abandonment of full employment policies and a focus on reducing inflation through restricting money supply were part of the reason for rejecting incomes policies, they were also rejected as being 'dangerous courses of action which were guaranteed to bring about a major, possibly fatal, clash between government and trade unions' (Dorey 1993:34). As such, the abandonment of incomes policies was a further means of attempting to reduce any political role for trade unions. Another important means of reducing the role of trade unions was to be through legislation.

The Legislative Framework

Until 1971, the primary role of legislation in regard to the activities of trade unions was to provide immunities from the common law which otherwise governed relations between trade unions and management (Morris and Archer 1992). The result of this system of regulation of industrial relations was a so-called voluntarist system of collective bargaining where the parties to industrial relations were free to make bargained agreements. There was no external authority regulating either the content or the negotiation of workplace agreements.

In the 1960s, concerns over the perceived extensive influence of trade unions in labour relations developed and a Royal Commission on Trade Unions and Employers' Associations (1965–68), referred to as the Donovan Commission, made a number of recommendations for the reform of industrial relations. While the recommendations of the Royal Commission were not adopted, the Conservative Party government, under the prime ministership of Edward Heath, did act to alter the relations between employers and unions by enacting the *Industrial Relations Act 1971*. According to Morris and Archer, the Act 'aimed to encourage legally enforceable collective agreements, to reform collective bargaining, to make unions liable for controlling and disciplining their members, to give union members greater rights against their union and to restrict industrial action' (1993:3). Neither unions nor the majority of employers supported the use of law in industrial relations at this time, and they succeeded in obstructing implementation of the Act.

By the late 1970s, a number of legal supports for collective bargaining had developed. These legislative enactments were based on the understanding that legal regulation of the employee–management relationship and the contract of employment required modification of the common law of commercial contracts, torts and criminal law. The election to office of a Labour government, under the prime ministership of Harold Wilson, in 1974, resulted in the repeal of the 1971 Act and the establishment of a 'social contract' involving the government and trade unions. In 1974, the *Trade Union and Labour Relations Act* was enacted which established the presumption that collective agreements were not legally enforceable unless expressly stated. A number of immunities from tort were established, and there were extensions of existing immunities. The *Employment Protection Act 1975* was enacted primarily to support and extend collective bargaining, while also creating a number of institutions including the Advisory, Conciliation and Arbitration Service (ACAS). Institutions such as ACAS represented an attempt by the Labour Party government to provide a third party which could be called upon to assist in disputes between employers and trade unions. As such, the development represented an important departure for the regulation of workplace relations in Britain in so far as it 'infringed' on what had previously been considered to be the sole domain of employers and worker representatives.

The electoral success of the Conservative Party in May 1979 marked the beginning of an era in which many of the supports for collective bargaining instituted during the 1970s were withdrawn by the government. For example, the *Employment Act 1980* removed the right of trade unions to be recognised as the representatives of the workforce in collective bargaining, and also removed the procedure for the extension of collective agreements. In 1983, the Fair Wages Resolution of the House of Commons, which obliged government contractors to observe terms and conditions not less favourable than those established by collective bargaining, was rescinded. Wages Councils, established in 1909 to provide statutory minimum wages and conditions in areas in which collective bargaining was weak, were abolished in 1993 (*Labour Research* 1993).

In addition to weakening the legal protection of collective bargaining, legislation in the 1980s limited other areas of trade union activity:

- The *Employment Act 1980* removed or restricted the immunities of trade unions in relation to certain forms of trade union activity, including the imposition of restrictions on the immunity for 'secondary' industrial action. It also restricted the ability of trade unions to enforce closed shops.
- The *Employment Act 1982* further tightened the restrictions on closed shops, restricted political strikes, and forbade union-only labour contracts.
- The *Trade Union Act 1984* restricted trade union funding of political parties, and also required postal ballots for strikes.
- The *Employment Act 1988* modified some of these enactments, and further regulated trade union activity through provisions relating to members 'unjustifiably disciplined' by their union. It also established a Commissioner for the Rights of Trade Union members.

These reforms, aimed at reducing the influence of unions on workplace relations, were extended further during the 1990s through legislation, including the *Employment Act 1990*, the *Trade Union and Labour Relations (Consolidation) Act 1992* and the *Trade Union Reform and Employment Rights Act 1993* (Morris and Archer 1992). The goal of such institutional reform was to free business from the perceived constraints to successful and profitable enterprise posed by the union movement, and to elevate individual workplace relations between employer and employee ahead of the collective representation of the interests of the workforce. Somewhat ironically, the government adopted an approach of non-intervention in the market at the same time that it instituted an increasingly interventionist role in British industrial relations, especially in regard to the internal regulation of trade unions (Ewing 1993).

The limiting of the scope of collective bargaining by the state made it increasingly straightforward for management to introduce new workplace practices. These practices included the deployment, organisation, and discipline of the workforce on a unilateral basis. Evidence of workplace reorganisation suggests, for example, that 'management's approach to the introduction of new technology [in the 1980s] was essentially opportunistic, consulting with the unions when it felt constrained to do so, but not otherwise' (Edwards et al. 1992:28). While workplace reorganisation was primarily management driven, the state played a crucial role in preparing the context for such reorganisation, often involving the introduction of technologies, by removing or restricting legal supports for collective bargaining. Such institutional reform was driven by a neo-liberal free market ideology in which the rights of the individual were paramount. The reform represented a major and fundamental reorganisation of the relations between business, the workforce and the state that was to affect the operation of individual workplaces.

The Role of Business

The 1980s also witnessed major changes in the practice and organisation of business in Britain, with important consequences for the introduction of technology into the workplace and for restructuring of workplace relations. An extremely active government which supported private business and the weakening of trade unions provided an institutional context that enabled business to introduce workplace reorganisation without necessarily consulting unions through collective bargaining.

A significant development in business practice directed at trade unions was union de-recognition, which is a practice whereby management makes a decision to exclude unions from the workplace. Although there is debate as to how widespread the practice of union de-recognition has become, it is acknowledged that the practice is particularly prevalent in both the national and provincial newspaper sector (Ewing 1993; McLoughlin and Gourlay 1994; Purcell 1993). De-recognition can be seen as one aspect of a more general trend towards union exclusion in the 1980s which also involved substantial contraction in the coverage of collective bargaining, a narrowing of the scope of collective bargaining, a reduction in the depth of union involvement in the workplace, and the erosion of the organisational security of trade unions (Ewing 1993). De-recognition also occurred in a climate where negotiation of workplace terms and conditions in many

industries shifted from multi-employer national agreements to negotiations occurring, if at all, at the local workplace level. In some
instances, no formal bargaining occurred at all, and employers simply
presented employees with terms and conditions of employment on a
take-it-or-leave-it basis (Millward et al. 1992; Purcell 1993). As such,
union de-recognition occurred in a broader social and political
context of transformation of workplace relations, linked to both
Conservative government policy and to a reorientation in business
practice.

Trade Unions and Technology

In the 1980s, British organised labour 'suffered its largest ever sustained
decline in membership' (Metcalf 1990:32). From 1979 to 1989, the
Trades Union Congress (TUC) lost almost 4 million members as its
membership dropped from 12.2 million to 8.4 million. By 1998,
membership of the TUC was 6.7 million, organised in more than
seventy unions (TUC 1998). While factors such as the generally
depressed economic climate, with high unemployment levels and shifts
in the job market from manufacturing to services, influenced membership levels, the legislative policy of the government was also important.
In particular, the legislation enacted in the 1980s made it easier for
employers to make decisions without entering into collective
bargaining with trade unions (Ferner and Hyman 1992; Purcell 1993).
A major survey of workplace relations from the 1980s supports this
claim, showing that from 1984 to 1990, there was a fall in aggregate
coverage of workplaces by collective bargaining agreements from 71
per cent to 54 per cent, with private sector coverage falling from 52
per cent to 41 per cent (Millward et al. 1992). As McLoughlin and
Gourlay observed, '[B]y 1990 a minority of employees in the private
sector had their pay and conditions determined by collective
bargaining' (1994:14).

A corollary to this decline in membership, and a diminished union
role in workplace bargaining, was a reduced input into government
and business policy decision-making by trade unions. Indeed, TUC
policy in regard to economic restructuring was effectively ignored by
the Conservative Party. While the trade union movement had achieved
some degree of legitimacy in the postwar period as an important
economic actor, the Conservative Party denied them any legitimacy,
claiming they were unrepresentative of the ordinary working person.
In pursuing this line, the government was also assisted by declining
union membership numbers, even where that decline could be linked
to government policy. Nevertheless, the TUC and various individual

unions continued to develop policy in regard to economic restructuring, and more specifically, in relation to technological development.

Through the tenure of the Conservative Party government, the TUC
developed policy regarding the introduction of technology into the
workplace. In *Employment and Technology* (TUC 1979), the TUC recognised the need to move beyond bargaining over employment issues,
and to consider various control and strategic issues involved in workplace relations (McLoughlin and Clark 1994). This report proposed
new technology agreements (NTAs) as a means of negotiating with
employers over the introduction of new technology into the workplace.
The TUC developed a checklist of points which were to be covered in
such agreements. These included the principles that change must be
by agreement, guarantees of job security, redeployment and relocation
agreements must be achieved, and innovation must result in improvements in terms and conditions of employment (Markey 1984;
McLoughlin and Clark 1994; Willman 1986).

Through the development of such agreements it was proposed that
trade unions would support technological developments in the workplace while also having an input at all stages of the process to minimise
the adverse consequences of these developments for employees. For
this policy to have succeeded it was essential that the TUC and the
union movement in general have the support of a sympathetic government. With the election of the Conservative Party to office in 1979,
such support was lost and by the late 1980s there was an acknowledge-
ment that NTAs were not a success (McLoughlin and Clark 1994; Noon
1991; Willman 1986). An attempt to reach an agreement in 1981 with
the Confederation of British Industry (CBI) over the processes of technological innovation in the workplace was also unsuccessful, with the
CBI rejecting the plan as interfering with established practices at firm
and industry level (McLoughlin and Clark 1994; Willman 1986). In
short, while the TUC developed various policies regarding technology
development within firms and industries, the political, institutional and
social climate of the 1980s and early 1990s was not conducive to the
implementation of such policies.

The societal and institutional context for technological innovation
and workplace reorganisation in Britain in the 1980s and 1990s underwent a major transformation. Whereas the union movement had
become increasingly involved at the highest levels of government in the
1970s, by the 1990s it had been marginalised. Government policy
provided a context in which business was able to reassert its dominance
in the workplace, and in which the very right of unions to represent
the collective interests of the workforce was rejected at a policy level.
Such developments were to have a major impact on the relation

between technological innovation and workplace reorganisation across all industries in Britain, especially the newspaper industry.

The Context for the Restructuring of Industry: Australia

As in Britain, workplace relations in Australia underwent important transformations in the 1980s and 1990s. While the Australian Labor Party, which was in office for much of this period, was committed to reorganisation of the economy and faced many similar issues to those confronting the government in Britain, the process of transformation was to be quite different.

Historically reliant on agricultural and mineral exports, and foreign investment, the Australian economy has been particularly vulnerable to fluctuations in world economic activity. One means by which governments have sought to enhance living standards has been through protectionist policies which have included tariffs and other trade barriers. These barriers were developed to protect domestic manufacturing from overseas competition. Periods of stagflation and more restrictive practices in the global market for agricultural goods, however, resulted in deteriorating terms of trade with the price of exports declining relative to the cost of imports. Linked to global economic restructuring from the 1970s, this decline in the terms of trade when combined with a lack of investment by domestic and international business in the Australian economy raised questions about protectionist policies. Governments and other economic actors came under increasing pressure to improve the balance of payments, and to make Australian manufacturing internationally competitive by restructuring the economy. Economic restructuring in Australia throughout the 1980s and into the 1990s involved the partial deregulation of the economy, increasing the flexibility of labour markets, and adopting micro-economic reforms aimed at improving the performance of individual firms. A central element in the strategy of economic restructuring was the development of the Accord between the Australian Labor Party government and the trade unions under the leadership of the Australian Council of Trade Unions (Archer 1992; Bell 1992a, 1992b; Frenkel 1988; Head 1983; McEachern 1991; Probert 1993).

The Accord: A National Framework for Economic Restructuring

The electoral victory of the Australian Labor Party (ALP) at the federal level in 1983 brought with it the promise of a shift from the monetarist policies that had been adopted by the preceding conservative coalition government of the Liberal and National parties. It also brought a new government relationship with the trade unions which had been one of hostility under the previous government. In the lead-up to the 1983 election, the ALP opposition had sought agreement over broad national economic goals with trade unions and business organisations, and this attempt to build consensus was reinforced through the 1980s in a series of Accords between the ALP government and the peak trade union organisation, the Australian Council of Trade Unions (ACTU). A number of tripartite organisations, involving government, trade unions and business groups, were also established as part of the industrial policy of the government.

The initial *Statement of Accord by the Australian Labor Party and the Australian Council of Trade Unions Regarding Economic Policy* (ALP/ACTU 1983) was an attempt to create a wages policy that could be used to respond to serious problems confronting the Australian economy, including low and uncertain economic growth, unemployment, inflation and low investment. According to the ACTU, in a report written with the Trades Development Council:

> The Accord, from its inception, had an integrated set of goals involving: working towards full employment; cutting inflation; increasing economic growth; maintaining living standards and improving them over time; developing industry; making the tax system fairer; improving industrial relations and occupational health and safety; improving the social wage . . . ; and industrial democracy. In return for the government's commitments under The Accord to full employment and other supporting goals, the trade union movement agreed to cooperate in this overall strategy and not seek wage increases beyond those prescribed in an accepted set of Wage Fixing Principles, set up under a return to a centralised system of wage fixation (ACTU/TDC 1987:49).

The Accord document indicated specific ways in which these goals were to be achieved. Consultation between government, unions and employers over issues including new industrial relations legislation, new workplace relations and practices, technological investment and industrial development policy was considered crucial. Such consultation was to be achieved through planning mechanisms, including the establishment of a representative Economic Planning Advisory Council and the preparation of industry plans. The Accord was also intended to facilitate the introduction of new technologies and new workplace relations to replace those which were increasingly being considered to be inefficient (Creighton and Stewart 1990). As McEachern suggests, 'The broad goal sought by The Accord was to solve the problem of the recession by regulating wages and promoting employment as well as the restructuring of the economy' (1991:20; see also Glyn 1992; Stewart 1992). The Accord evolved through a number of modifications or stages in the 1980s, all of which 'dealt with wages policy and all, even those arguing for larger wage increases, have been framed to promote economic and employment growth' (McEachern 1992b:97).

An important element of the Accord was its focus on restructuring workplace relations in Australia, and on creating an institutional context which would facilitate technological innovation and investment. While the government and various state institutions and actors were to play a direct role in these processes of restructuring, much of their activity was focused on creating the conditions in which business, with the agreement of unions, would undertake reorganisation of the workplace.

The Industrial Relations Commission: A System of Conciliation and Arbitration

In contrast to the voluntarist system of industrial relations regulation that predominated in Britain, the state has played a major role in industrial and workplace relations in Australia throughout the twentieth century, in particular through the operation of the conciliation and arbitration system (Frenkel 1988; McEachern 1991). Under Section 51(xxxv) of the Australian Constitution, the federal government can make laws with respect to 'conciliation and arbitration for the prevention and settlement of disputes extending beyond the limits of any one state'. In terms of setting wages and conditions, the federal government can establish tribunals such as the Industrial Relations Commission, but then cannot 'fetter the [Industrial Relations] Commission's exercise of the wage fixing part of the machinery' because 'fixation of wages is exclusively a matter for the Commission's discretion' (Singleton 1992:131). While the federal government can present arguments to the Commission about wages and conditions of work, it cannot set them itself. The combination of wage-fixing functions and its involvement in the conciliation and arbitration of disputes over workplace conditions between employers and trade unions meant that the Industrial Relations Commission had an influential role in the restructuring of Australian industry in the 1980s and 1990s.

The Industrial Relations Commission has been central to the Australian industrial relations system in various forms since 1904. Developed initially in the aftermath of a series of strikes in the 1890s, with the recognition that the individual common law contract of employment was insufficient by itself to handle the collective nature of industrial and workplace relations, the Australian system of institutional collective dispute resolution has been one of compulsory conciliation and arbitration. The two fundamental features of the system are the use of a permanent and independent state-funded tribunal to exercise the functions of conciliation and arbitration, and the compulsory nature of the system. The parties to a workplace dispute may be compelled to submit their differences to the Commission for settlement, and any resolution imposed by or adopted under its auspices will be legally binding on the parties (Creighton and Stewart 1990). Workplace relations until 1996 at the federal level were governed by the *Conciliation and Arbitration Act (Cwlth) 1904* and its successor, the *Industrial Relations Act (Cwlth) 1988*, which recognised trade unions as the legitimate representatives of working people. The system also regulated the employment relationship through the provisions of awards, orders and certified agreements. Such forms of regulation were determined by industrial tribunals in hearings involving the relevant parties that regulated factors including wages, paid leave, the organisation of work relations, who does specific categories of work, and health and safety (Mitchell and Rimmer 1990).

In the 1980s, criticisms arose that the Commission was excessively legalistic and inflexible, and that it was contributing to the inefficient and non-competitive state of the Australian economy (Creighton and Stewart 1990). Sections of business, the media, and of the conservative political parties called for the abolition of the arbitration system, and for its replacement by direct negotiations between employers and employees. Other actors, particularly the federal government and trade unions under the auspices of the ACTU, sought workplace reform through the system (Stewart 1992). A way of reforming the workplace emerged through the arbitration system in the form of award restructuring. As the bargaining process between the government, trade unions, and business developed through the 1980s, a shift occurred in wage-fixing processes to forms of productivity bargaining. Workers were granted wage increases in return for accepting reorganisation of workplace relations intended to increase efficient production. There was also a shift towards enterprise bargaining, which involved direct negotiating at the enterprise level between employers and employee representatives. The combination of productivity and enterprise bargaining was developed as a major means of achieving microeconomic reform at a time when the centralised industrial relations system, and industry or sectoral bargaining, were considered to be too rigid and inflexible to meet the challenges of international competitiveness and economic recession (ALLR 1992). Such wage-fixing reforms were accepted in National Wage Case decisions of the Industrial Relations Commission, while the concept of enterprise bargaining was enacted into legislation as Section 170LC of the *Industrial Relations Act 1988* (Brooks 1992; Singleton 1992; Stewart 1992). The movement towards enterprise bargaining was further emphasised in 1996 with the passage of the *Workplace Relations Act 1996*, which replaced the *Industrial Relations Act 1988*, under the newly elected Liberal–National Party coalition government.

Throughout the 1980s and early 1990s, the Industrial Relations Commission was active in shaping the arguments of various other actors into potentially workable and 'politically acceptable' outcomes. The Commission rejected, for example, the concept of enterprise bargaining in the National Wage Case of April 1991, on the basis that the parties needed to clarify their objectives. However, in the National Wage Case of October 1991, after sustained argument from employers, the government and unions, the concept was accepted by the Commission (ALLR 1992; Singleton 1992). Significantly, the structure of the Commission meant that trade unions were active participants in the industrial reform process, while decisions of the Commission emphasised the need for agreements to be made between registered organisations of employers and employees. The *Workplace Relations Act 1996* suggests that this role for the Commission may well be diminished in the years ahead as the Liberal–National Party coalition government embraces the operation of individual bargaining in the workplace free from the perceived constraints of both the state and the collective organisation of workers.

Government Committees and Plans

While the Industrial Relations Commission provided an important forum for examining industry restructuring, and for facilitating technological innovation, the federal government undertook various initiatives itself to establish a context in which to assist these reforms. Among these interventions was the creation of the Economic Planning Advisory Committee, a tripartite consultative body involving the government, the ACTU and representatives of business. It was established to 'expand the information base available for economic policy formation through broad indicative planning' (McEachern 1991:66). The government also formed the Advisory Committee on Prices and Incomes, and the tripartite National Labour Consultative Council (Davis and Lansbury 1989). Another important organisation was the Australian Science and Technology Council (ASTC) which published reports including *Science, Technology, and Australia's Future* that identified 'ways in which science and technology may be used strategically to improve the growth prospects and sustainability of Australia's production of traded goods and services' (ASTC 1990:1). Industry plans were also developed, including strategies for the steel and car industries. The aim of the government was stated by the then Minister for Industry and Trade, Senator John Button, who claimed that substantial modifications to work practices were needed which would achieve improved efficiency, avoid demarcation disputes and broaden occupational classifications, as well as enable 'better management practices to be introduced, clearing the way for the introduction of new technologies and equipment' (Johnson 1989:106). Whether these industry plans and other government initiatives achieved their goals has been the subject of debate (Bell and Wanna 1992; Cook 1992; Evatt Foundation 1995; Johnson 1989; McEachern 1991). It is agreed, however, that the government attempted to create a context in which restructuring of the national economy and of workplace relations and practices could occur based on consensus and participatory agreement, rather than on conflict and confrontation. The establishment of such an institutional and societal context provides a stark contrast with the British experience over the same period where the rights of the individual were promoted, and where the government sought to minimise its direct role, and the role of unions, in relation to workplace bargaining over technological innovation and workplace reorganisation.

Trade Unions and Strategic Unionism

Throughout this period of industrial reform in Australia, the role of trade unions was particularly significant. In addition to presenting arguments to the Industrial Relations Commission in National Wage Cases, the ACTU developed increasingly sophisticated policies regarding the introduction of new technologies into the workplace through 'strategic unionism'. Strategic unionism was defined by the ACTU as a policy which would allow unions to 'go beyond a narrow focus on wages and conditions', and generate and implement 'centrally coordinated goals and integrated strategies; e.g. for full employment, labour market programs, trade and industry policy, productivity, industrial democracy, social welfare, and taxation policies which promote equity and social cohesion' (ACTU/TDC 1987:169). Among the ways in which the ACTU sought to pursue such goals were participation in tripartite bodies, an emphasis on strong local and workplace organisation, and delivery of education and research services (ACTU/TDC 1987). Strategic unionism involved trade unions not only pursuing industrial goals but also being active in the political arena. The result was the development of ACTU policy that was concerned with wide-ranging workplace reorganisation. At its 1985 Congress, for example, the ACTU encouraged its affiliates to negotiate with employers on matters including technology, job security, occupational health and safety and job design. In September 1986 the ACTU, the Confederation of Australian Industry and the Business Council of Australia signed a joint statement which stated that 'the parties acknowledge the need for developing more effective employee participation based on improved information sharing, developing more effective communications between all levels of the enterprise and encouraging more active employee involvement' (Lansbury and Davis 1992:233).

The policy of strategic unionism was developed in the context of the Accord, and was assisted by the consultative emphasis of the Accord. Strategic unionism and the Accord also strengthened links between the Labor government and the ACTU which resulted in union input into government policy formation. The movement of individuals such as Bob Hawke and Simon Crean from positions of leadership in the ACTU to being prime minister and a senior cabinet minister respectively was important, while other union leaders such as Bill Kelty and Laurie Carmichael were influential in bringing union concerns to the attention of the government (Carney 1988). Again, such experiences provide a very different setting from Britain where unions were explicitly excluded from any participation in government policy formation in the 1980s and early 1990s, and where unions were seen as fetters on the operation of the market rather than as social partners with a legitimate role in policy formation.

Business and Opposition Political Parties

An important element in the development of the Accord in the 1980s was the relative disunity of business and of the Liberal–National Party opposition (McEachern 1991). Such disunity was illustrated dramatically at the 1988 National Wage Case where 'ten separate employer submissions were made to the Arbitration Commission ranging from opposition to the very existence of the arbitration system to support for the ACTU's wage claim' (Archer 1992:411). Disunity was also illustrated with the establishment of the Business Council of Australia (BCA) by a number of Australia's largest corporations, which were frustrated with their perceived lack of representation in the Confederation of Australian Industry (CAI), whose membership was restricted to business associations (Frenkel 1988). The relationship between business and the Liberal–National Party opposition – the traditional political representative of business – was also subject to strain in the 1980s. In particular, the emergence of the New Right in organisations such as the HR Nicholls Society and the Australian Federation of Employers represented a radical market-oriented critique of the traditional conservative elements of the Liberal Party, as well as of the Labor Party (McEachern 1992a, 1992b). As a result of such divisions, differences emerged within the opposition forces on the best way to counter the Labor government policies and on how to promote electorally appealing alternatives. One of the consequences of such divisions was that the conservative political parties were not to regain federal office until 1996.

The Context for Industrial Reform: An Example

A case brought to the Australian Conciliation and Arbitration Commission (which became the Australian Industrial Relations Commission in 1988) in 1983 further illustrates the political relations and alliances established in the 1980s, which provided the context for industrial reform. In the *Termination, Change, and Redundancy* (TCR 1984) case, a number of trade unions represented by the ACTU brought claims seeking 'significant improvements in three main situations: firstly, on termination of employment, secondly, on the introduction of change by an enterprise, and thirdly, in the event of redundancy' (TCR 1984:177). In relation to the introduction of change, the ACTU argued for the establishment in federal awards of 'obligations on employers to notify and consult with employees about the introduction of new technology' (TCR 1984:177). According to the Commission, the primary argument of the ACTU was that 'the opportunity to discuss matters such as job requirements, training, job security, working hours, monitoring the change and so on would minimise the potential for conflict which exists when changes are introduced with significant benefits for industrial relations' (TCR 1984:195). The ACTU received the support of the federal Labor government, as well as of four state Labor governments, for the inclusion in federal awards of 'improved standards in relation to . . . obligations on employers to consult with their employees and their unions on production technology and other changes likely to have significant effects on employees' (TCR 1984:179).

In contrast, the CAI, with the support of conservative governments in the two other states, supported the principle of consultation, but argued for a voluntary approach. Their case was based on the argument that a voluntary approach would allow management to retain flexibility in making workplace decisions, while also maintaining responsibility for issues including timing, content and implementation of innovation and reorganisation. The CAI 'also objected to the widespread nature of the changes covered by the claim and the delay that would be caused to an employer seeking to implement change' (TCR 1984:195). It suggested that 'the clause could be used by unions who have a fundamental and long-standing objection to technological change to frustrate the implementation of change' (TCR 1984:195). That is, the CAI argued that these issues did not relate to the terms and conditions of employment, but rather were issues about the function and role of management and, as such, should not be the subject of a legally binding award.

The Commission was hesitant to make an award including require-ments about procedures for negotiation, consultation and provision of information on the grounds that these were not matters that effectively lent themselves to such awards. It was prepared, however:

> to include in an award a requirement that consultation take place with employees and their representatives as soon as a firm decision has been taken about major changes in production, program organisation, structure or tech-nology which are likely to have significant effects on employees.
>
> We have decided also that the employer shall provide in writing to the employees concerned and their representatives all relevant information about the nature of the changes proposed, the expected effect of the changes on employees and any other matters likely to affect employees. However, we will not require an employer to disclose confidential information (TCR 1984:196).

Through decisions such as this, the Commission played an important role in creating a supportive context for the implementation of the Accord in the workplace. The emphasis on negotiation and consulta-tion over workplace reorganisation is apparent, while the decision also recognised that employers ultimately control the investment process. In addition, the case highlights the alliances that were built up around the issue of technological innovation, and the arguments for and against consultation over such processes. By seeking voluntary consul-tation, the CAI was effectively arguing for unions to be granted an unenforceable right. The claim by the CAI that consultation regarding technology was not an industrial matter was an attempt to put it beyond the jurisdiction of the Commission, and thereby avoid the enforcement capacity of the Commission. This case illustrates how various industrial, legal and political actors influenced the relationship between techno-logical innovation and workplace reorganisation in Australian industry in a manner very different from the process that was developing simul-taneously in Britain.

In summary, the 1980s in Australia witnessed investment in tech-nology associated with major workplace reorganisation. Typically, the association was mediated by interactions between management and owners, trade unions and workers, and influenced by various state institutions and state actors, as well as by general societal conditions

(Callus et al. 1991). The Accord was to be the basis for the creation of an industrial context that would promote productive investment and other forms of economic activity which, it was hoped, would improve the international competitiveness of the Australian economy. As part of this process, the Accord sought to establish 'the necessary environmental conditions for a major reorientation' in the structure of workplace relations (Arup 1991:53). The processes developed for such a reorientation involved bargaining, negotiation and agreement between institutions of the state, labour and capital. There was an attempt in the 1980s, at least at a formal level, to pursue reform through negotiation and consultation between the key actors in the economy. The establishment of such an institutional and societal context contrasted with developments in Britain in the 1980s and into the 1990s where business was seen as the appropriate decision-maker in the economy, with the role of organised labour being reduced to the narrowest scope possible. The Australian experience was also to provide a striking contrast to developments in the US from the late 1970s to the 1990s.

The Context for the Restructuring of Industry: The United States

For much of the twentieth century, the United States has been regarded as having the world's strongest economy. In 1913, for example, the US was the largest producer of electricity, cars and oil, while by 1926 the US produced 45 per cent of the world's industrial output (Marshall 1994). Following the Great Depression of the 1930s, the post-World War Two years were marked by 'unanticipated abundance' (Bluestone and Bluestone 1992:34), as the American middle class expanded on the basis of unprecedented private sector investment and government spending and provision of infrastructure. In addition, the US became the globally dominant economy, benefiting from its postwar investment in the Truman Doctrine and the Marshall Plan.

By the 1970s, however, the US economy had entered a period of crisis, characterised by declines in economic activity in areas such as productivity, quality and innovation (Bluestone and Bluestone 1992). In the wake of the oil crisis of the early 1970s, the loss of competitiveness of the American economy on a global scale became a major concern as profits declined and standards of living were threatened.

Reaganomics

As in Britain, the most notable governmental response to the crisis marking the decline of the Golden Age of Capitalism in the US came with a neo-liberal administration seeking to free the market from the perceived constraint of the state and organised labour. While a turn to neo-liberal policies was presaged by the policies of various administrations in the 1970s, they came to fruition under the presidency of Ronald Reagan, elected in 1980, and were to a large extent continued by his successor George Bush who was elected president in 1988. According to the economists Bowles, Gordon and Weisskopf (1990), the key elements of the government policies introduced in the 1980s were:

- using tight monetary policy to generate high interest rates;
- attacking labour unions and intimidating workers;
- promoting government deregulation of business;
- using tax policy to shift income to the wealthy and corporations;
- regaining international economic advantage through remilitarisation.

As with the Thatcher government in Britain, government policy in the United States in the 1980s and into the 1990s under presidents Reagan and Bush was based on the ideology that the best way to rejuvenate the economy was to allow individuals and businesses to use their initiative in the market, independent of interference from government regulation. Deregulation was central to this strategy and involved, for example, refusal to implement key environmental regulations aimed at decreasing pollution (Ferguson and Rogers 1986); liberalisation of banking; and lifting of price controls in the long-distance telecommunications market (Boyer 1991). In addition, the government made taxation a major policy, sharply reducing personal income tax rates and providing business with new tax breaks (Pierson 1994). As with the Thatcher government in Britain, unions and the institutions of collective bargaining were also identified as key fetters on the operation of the market. In the early years of the Reagan administration, for example, the dismissal of all 11,345 members of the Professional Air Traffic Controllers' Organization (PATCO) signalled the intention of the administration to deal severely with unions that were considered to be obstructing the operation of the market.

As in Britain, the approach of US administrations in the 1980s towards technology was to seek to provide a context in which businesses could successfully invest in technologies. Rather than providing direct assistance for developments in technology, apart from the military and national defence, the administration considered that its role was to renew the entrepreneurial spirit in the US through tax breaks, reinforced patent legislation and the repeal of antitrust restrictions (Derian 1990). President Reagan's faith in the ability of business to provide leadership in technological innovation was captured in his State of the Union address in 1982, when he commented that the pioneering spirit that had made the US the major industrial actor in the twentieth century was now operating to open up the opportunities of the high technology frontier (Derian 1990).

As in Britain, a realignment of the relations between the state and society was occurring in which private business interests were considered to be the most important actors in society. Business was to be allowed to operate independently of constraint, whether from the state, unions or other actors. John Sununu, the White House Chief of Staff during the Bush presidency, captured the essence of this perspective with his observation that the President believed that the free market operated best when not 'shackled' by government (Nester 1998:206).

While there has been debate over the success of this deregulatory strategy (Nester 1998; Pierson 1994; Sahu and Tracy 1991), there is little doubt that organised labour felt the brunt of neo-liberal policies in the US (*Guild Reporter* 1994b). One area in which such policies were to affect labour was in the legal regulation of workplace relations.

Legislative Regulation of Workplace Relations

The basic legislation governing workplace relations in the US in the twentieth century was enacted in the New Deal era of the 1930s, and was to become the basis of the New Deal industrial relations system (Kochan, Katz and McKersie 1994). This legislation sought to provide some state regulation of workplace relations in the wake of the ravages which had befallen the workforce in the years of great depression. In particular, the Democratic administration, under the presidency of Franklin Roosevelt, accepted the need for some form of social regulation to protect workers' rights; to achieve a balance in the power relations in the workplace; and to accommodate the diverse interests that existed in the workplace (Kochan, Katz and McKersie 1994). Among the key Acts passed in this period were the *Davis-Bacon Act 1931*; the *Anti-Injunction (Norris–LaGuardia) Act (1932)*; the *National Industrial Recovery Act 1933*, which was to be declared unconstitutional by the Supreme Court in 1935; the *National Labor Relations (Wagner) Act 1935* which replaced the 1933 Act; the *Walsh–Healy Act 1936*; and the *Fair Labor Standards Act 1938* (Knoke et al. 1996). In addition, this legislatively enacted framework was also supplemented by common law decisions of the court system (Cox et al. 1991; Tomlins and King 1992; Weiler 1990). The basic law governing union–management relations was the *National Labor Relations Act 1935* (NLRA), while the National Labor Relations Board (NLRB) was also created to regulate and enforce federal labour legislation (Weiler 1990). Section 7 of the NLRA of 1935 provides in part that:

Employees shall have the right to self-organisation, to form, join, or assist labour organisations, to bargain collectively through representatives of their own choosing, and to engage in other concerted activities for the purpose of collective bargaining or other mutual aid or protection.

Collective bargaining was fundamental to the New Deal system of industrial relations, with unions recognised as the legitimate representatives of the workforce in situations where employees chose to be so represented. When a union wins recognition rights in the workplace through ballots, the employer is statutorily required to bargain with the union on wages, hours and other terms and conditions of employment. As in other common law jurisdictions, the scope of the concept of terms and conditions of employment has been subject to varying legal interpretation. Initially, all bargaining was to be left to the parties, with no outside interference from courts or other legal bodies. In the years following the enactment of the New Deal legislation, however, the NLRB and the courts began to clarify which issues were the proper subject of workplace bargaining by distinguishing between mandatory, permissive and illegal subjects of bargaining. Issues which have been interpreted as being the proper subjects for collective bargaining include seniority, layoff and recall rules, discharge and discipline provisions, grievance and arbitration procedures, no-strike and lock-out clauses, and union security arrangements, in addition to health and safety matters (Bellace 1992). Crucially, however, management was to retain its prerogative over strategic business decisions. One issue not included directly as a proper subject of collective bargaining is new technology. As Bellace observed:

In the United States, labour lawyers tend not to discuss the subject of new technology in those terms. Certainly, new technology is introduced into the workplace. Either the workers are receptive to it and no labour law issue is created, or the workers reject the way in which management proposes to introduce it. For instance, management may state that new equipment will be introduced, thus eliminating the need for half the workforce. Management may also say that with the new equipment, new job classifications with new wage rates are needed. The union can do nothing about the reduction in the workforce . . . Because job classifications and wage rates are a mandatory subject of bargaining, the union can insist to the point of impasse over these items.

Hence, with regard to the introduction of technology, it can be said that American unions break down the overall phenomenon into its component parts and then determine which items are subject to mandatory bargaining (1992:249).

That is, the introduction of technology into the workplace, and its consequences for the workforce, can only indirectly be the subject of legally mandatory workplace bargaining and negotiation.

Despite the general acceptance of collective bargaining in the years following the New Deal legislation, various other Acts were introduced which limited the scope of collective bargaining. The *Labor Management Relations (Taft–Hartley) Act 1947*, for example, outlawed secondary boycotts, made the closed shop illegal, allowed states under Paragraph 14b to outlaw the union shop, and allowed employers to bring unfair practice claims against union representatives (Bluestone and Bluestone 1992; McCammon 1990; Rogers 1990). In addition, the *Labor–Management Reporting and Disclosure (Landrum–Griffin) Act 1959* legalised intervention into union affairs by government officials to a greater extent than had previously been the case (Goldfield 1987). While such Acts did limit the scope of collective bargaining, for much of the postwar period collective bargaining was accepted as legitimate, providing unions with a powerful institutional basis on which to expand their membership, and to influence workplace relations.

With the economic crisis of the 1970s, however, this legal and administrative support for collective bargaining was challenged. In the 1980s, this took the form at an administration level of a lack of support for labour. Under the conservative leadership, 'legislation favouring labour . . . had a difficult time reaching the national agenda' (Knoke et al. 1996:123). One instance in which a pro-labour issue did reach the national agenda was a 1988 trade bill passed by the House and the Senate 'requiring companies with more than 100 employees to give at least 60 days' notice before closing plant or laying off at least one-third of its workforce' (Knoke et al. 1996:125). This bill was vetoed by Reagan, and was only enacted after Congress overrode the President's veto in the face of strong popular support. While the administration was hostile towards organised labour, many of the specific challenges to collective bargaining from the state in this period came from the legal sector. Commenting on judicial and administrative developments in the 1980s, Davis wrote:

> The principal state and judicial instruments of labour policy – the Supreme Court, the NLRB, and the administrative and regulatory agencies of the Department of Labor – have been motivated to reverse thirty-five years of policy and accepted procedure. It is not just that the administration has struck sharp blows against labour . . . but that it has created an executive and judicial framework most conducive to the new accumulation strategies of capital. By partially 'deregulating' labour relations along with banking and transportation, Reagan has paved the way for more rapid capital flight, plant closure, deunionisation, and the proliferation of all manner of new sweated industries (1986:138–9).

In addition, in the 1970s and 1980s there was important regulatory activity at the state level, through state legislatures and through state court judges, especially in the area of wrongful dismissal which had potential ramifications for those made redundant through technological innovation (Weiler 1990).

Crucial also to the changing institutional and legal regulation of labour relations in the 1980s was the conservative 'free market' political persuasion of new appointees to the National Labor Relations Board (NLRB). With a new Chair in Donald Dalston, the NLRB overturned forty of its own doctrines relating to collective bargaining and workplace relations, and developed a strong anti-union position with the intention of removing perceived legal constraints on the power of management in the workplace. A celebrated example of this approach to labour relations arose in the Milwaukee Spring case in 1984, which reversed earlier NLRB decisions by finding 'that employers need not bargain over plant closure and relocation to non-union sites if the closure does not "turn" on labour costs' (Davis 1986:140). This decision was supplemented by further rulings of the NLRB which 'allowed employers to contract out work without bargaining with unions, and supported their right to reopen contracts to demand concessions' (Davis 1986:140). As in Britain, these actions were accompanied by 'developing and expanding intrusive regulatory constraints on trade union action' (Weiler 1990:19). Finally, whereas under the administrations of presidents Ford and Carter in the 1970s the NLRB decided 28 per cent of unfair labour practice cases in favour of employers, in the 1980s 60 per cent of such decisions were in their favour (Katz and Kochan 1992). That is, institutions including the National Labor Relations Board, government bodies, and the Supreme Court influenced workplace practice in the 1970s and 1980s, and they were especially important in assisting in the creation of a context in which the prerogatives of management and employers in the workplace were considered paramount.

Labour–Management Relations

In relation to bargaining over and implementation of workplace reorganisation influenced by technological innovation, the political, social and economic relations of the United States in the 1970s and 1980s were unfavourable to trade unions, and supportive of a reassertion of the so-called right of management to manage. This period represented a decline in an accord that was achieved between employers and various unions in post-World War Two United States, and was to solidify the emergence of a non-union industrial relations system (Kochan, Katz and McKersie 1994; Rogers 1992). Building on the New Deal industrial relations legal system, the postwar accord represented an unofficial agreement between employer and unions in support of collective bargaining. In the manufacturing and construction sectors, in particular, unions achieved sufficient levels of membership density to take wages out of competition, 'thus removing employers' chief reason for resisting unionization' (Rogers 1992:290). Other factors, including the strong market position of leading manufacturing firms, strong domestic demand, and a relatively closed market, meant that in many cases firms could pass on increased costs attributed to unionisation to consumers. Within the core sectors of the economy, a relative degree of cooperation and stability in employer–union relations was achieved, accompanied by the development of a typical workplace contract. A typical workplace contract under the postwar accord contained seven basic elements:

- a specific wage fixation formula tied to productivity increases;
- job-related benefits attained through collective bargaining;
- seniority-based lay-offs, recalls and transfers;
- negotiated conditions of work, work rules and job classifications;
- grievance and arbitration machinery to resolve disputes over the contract;
- on-site union representation and a union security provision; and
- retention of management prerogatives over many workplace issues and virtually all strategic enterprise decisions (Bluestone and Bluestone 1992; Rogers 1992).

For much of the post-World War Two period, this arrangement worked to the benefit of both employers and unionised workers. Indeed, by 1953 union membership in the US reached a high point with 35 per cent of the workforce in unions (Davis 1986; Goldfield 1987).

Unfavourable economic developments in the early 1970s, however, resulted in concerted challenges to this period of relative workplace stability, and to the continuation of the accord. Declining economic performances at the national level, sharp drops in profits and Gross National Product, and declining or stagnating median family income and weekly earnings, when considered in conjunction with the internationalisation of the US economy, contributed to increasing variation in workplace relations. In particular, firms became more sensitive to costs, including the price of labour. This led to confrontations with organised labour, as the removal of unions from the workforce was considered by employers to be one means of reducing costs. As a corollary, and also in response to increased competition and uncertain demand, firms attempted to introduce increasingly flexible practices of production, whereby workers were to perform many and varied tasks in the workplace. These moves resulted in more confrontation between management and unions, and in some instances there were attempts by management to remove unions from the workplace, with unions considered to be protective of demarcations between jobs, and therefore impediments to flexibility in production. What emerged in such contexts was a system of non-union human resource management, in which individual workers and work groups were promoted at the expense of unions and collective organisation of the workforce. This strategy was also to manifest itself in the 1980s with a number of firms across the US closing down unionised plants and shifting to new non-union worksites, while union decertification elections were increasingly prevalent and successful (Kochan, Katz and McKersie 1994; Weinstein and Kochan 1995). These developments coincided with the creation of new technologies, the diffusion of which 'permitted firms to shift production away from externally organised markets – thus overcoming the locational immobilities that had benefited labour – or to break union monopolies on the generation of needed skills by organising alternative sources of skill, or deskilling the relevant production processes' (Rogers 1992:292–3). Examples of such technological developments included numerically-controlled machine tools in machining, containerisation in longshore work, and computerised composition and production in the printing and newspaper industry. The combination of new technology and changing economic circumstances were important in the breakdown of the accord between labour and management that had existed since World War Two (Kochan, Katz and McKersie 1986; Moody 1988).

New Technology and Workplace Relations in a New Industrial Context

Within this new industrial and institutional context, relations between management, unions and the workforce, particularly since the 1970s, have undergone major developments on the shopfloor. Accompanying the decline in union membership from its peak in the 1950s to a situation in the 1990s where union membership is less than half of that peak has been the emergence of a non-union industrial relations system as noted above, in which employers actively avoid union representation in the workplace (Deutsch 1993; Kochan, Katz and McKersie 1994).

Another important development in labour–management relations has been a notable rise in episodes of hostility and conflict with the breakdown of the postwar accord. This is attributable, in part, to the associated development of new labour-saving computer technology, and the determination of many employers and managers within an evolving industrial context to make use of that technology with or without the consent of the workforce. While instances of violence are reported, other tactics have become increasingly prominent in the reorganisation of the workplace by management, including the almost two-fold increase in unfair labour charges brought against employers in the NLRB during the 1970s; the growing number of election delays by employers in holding certification elections; and the increasing number of union decertification elections (Goldfield 1987). Crucial to these attempts to remove unions as an influence in the workplace has been the development of professional management consulting firms, also referred to as union-busting organisations. The role of such organisations was stated in a 1979 hearing on Pressures in Today's Workplace by the House Sub-Committee on Labor–Management Relations. The sub-committee reported:

> These [management consulting] firms provide a variety of services which might include union 'prevention', management and supervisor training, devising employee compensation programs, and other functions aimed at structuring the relationship between employer and employee to maximize employer control and minimise union influence. Their primary function, however, is the orchestration of campaigns to defeat unions (SOGAT 1985:15).

Writing in 1978 as an associate director of the National News Council, and as a former chief labour correspondent of the *New York Times*, A.H. Raskin was also aware of the growth of a more aggressive element in management practice, especially in dealings with unions. He wrote that, with the decline in union membership, falling hopes for labour law reform, and the loss by labour of bargaining units and workplace rights, management 'smells blood' (1978:197). He claimed:

> To an extent without parallel since the violent struggles of the early New Deal, when major companies built up private armies of goons and labour spies to smash infant unions in the mass production industries, employers are on the attack in labour relations. True, they no longer use tear gas, brass knuckles or guns, nor is there any broadscale offensive in the union heartland of the Northeast and Middle West . . . Now the tactics are more subtle and more selective. A whole new industry has grown up of 'labour relations consultants' specialising in union-busting sugarcoated as industrial psychology and behavioural science (1978:197).

Raskin noted that these developments signalled a new toughness had developed in the approach of management to collective bargaining negotiations with unions, which was to prevail in many industries including the newspaper industry.

Negotiations between management and unions in the 1970s and 1980s were also characterised by the extent to which management succeeded in gaining concessions from unions. Linked to high unemployment levels, economic recession and technological development, managements were able to reduce job losses by demanding 'give-backs' of concessions including wage cuts; ending automatic indexing of wages to the retail price index; reducing holidays and pension rights; no-strike agreements; and two-tier wage schemes (Upham 1993). In the newspaper industry, concessions were won from unions by management at newspapers including the *Boston Herald–American*, the *Oakland Tribune*, the *New York Post*, and the New York *Daily News* (Mitchell 1994). Technological developments were important in this regard in that they provided management with a vital tool in negotiating with unions. When combined with high unemployment levels, the ability of management to threaten large-scale lay-offs by introducing new technology severely weakened the bargaining position of the unions. Management was assisted further in this process by a ruling of the National Labor Relations Board to the effect that employers could continue to operate business lawfully by hiring temporary replacement workers in economic disputes, provided there was no specific evidence of anti-unionism. The bargaining position of management in strike situations was further strengthened by a decision of the Supreme Court in 1985 to uphold a ruling of the NLRB to the effect that unions could not discipline members who resigned during a strike (Upham 1993). These decisions also had the effect of weakening the bargaining position of unions, as it meant that in many instances the withdrawal of labour would not result in the cessation of production. Technological developments were again important in this context because in many instances the availability of computer technology meant that it was easier for employers to train or recruit new employees, especially in craft industries which had previously required highly skilled labour.

There are also examples of cooperative negotiation between management and unions over many issues during this period, however, including, in some instances, technological innovation. In arguing that there does exist a 'cooperative climate in which unions and management are jointly addressing problems at the workplace in a non-adversarial fashion', Deutsch and Schurman make the important point that the possibility of diversity in workplace relations and practices across the US needs to be recognised (1993:346). They cite the example of the agreement between the Saturn Division of General Motors and the United Automobile Workers 'as an exemplar of worker participation at all levels of planning and production, including technology and work organisation' (1993:349; see also Katz and Kochan

1992; for national figures on consultation clauses regarding technology in union contracts, see the Bureau of National Affairs, various editions). While the overall context for labour involvement in technological innovation and workplace reorganisation has not been favourable, such instances show that it is not impossible.

As the US economy experienced difficulties from the late 1960s, related to profitability, inflation and unemployment, so business and government began to consider alternative means of organising economic and workplace relations. While there is much debate concerning the overall success of the Reagan and Bush administrations in achieving their goal of deregulating the economy, there is little doubt that they succeeded in relation to organised labour.

Conclusion

These analyses of the institutional and societal contexts in Britain, Australia and the United States indicate the differences between the settings in which technological innovation and workplace reorganisation were being introduced. The relational model of workplace reorganisation indicates the need to understand the impact of such national contexts if variations in the relationships between technological innovation and workplace reorganisation in a particular industry are to be explained. It is the purpose of this study to use the relational model to examine the impact of new technologies in workplace relations on newspapers owned by News Corporation in Britain, Australia and the United States. Before that analysis relating to News Corporation is presented, the following chapter examines how the newspaper industries in the three countries reacted to the introduction of technological transformation, given the differing institutional and societal contexts that had been established.

CHAPTER 3

The Newspaper Industry: Historical Developments in the Three Countries

A number of studies have examined the influence of new technology on workplace relations in the newspaper industry. In addition to presenting the histories of particular newspapers and biographies of newspaper proprietors, these studies have focused on particular actors in the newspaper industry, including print and production workers and their unions; journalists and their unions; and management and proprietors.

Studies of print workers in the newspaper industry have emphasised the degree of control and influence that these workers and their unions have historically exercised in the workplace. Print unions have been able to control the supply of labour and skills in the industry through apprenticeships, and define their own jurisdiction in the face of opposition both from employers and other unions. Studies of the contemporary newspaper industry suggest, however, that developments in computer technology have transformed the skill requirements of the print process, resulting in a shift in the balance of power in the newspaper workplace from print unions to employers. In particular, computer technology has removed the need for craft skills, so that the work of a printer can be accomplished by any individual with basic computer literacy (Cockburn 1991; Cornfield 1992; Emery, Emery with Roberts 1996; Frenkel 1990; Griffin 1984; Kalleberg et al. 1987; Merrill 1995; Scott 1987; Zeitlin 1985).

Another group of studies has examined the influence of technology on the work of journalists, arguing that the consequences of technological developments for these workers may be very different than for print workers. Reed has suggested, for example, that:

> [c]omputerised (or automated) typesetting has the potential to deskill the printing trade, but it affects journalists in a different manner ... [W]here

journalists' claims of indeterminacy in day-to-day practice coincide with proprietary publication strategies there is scope for journalists to mobilise in support of enhanced control over the labour process and greater autonomy commensurate with professional status (1991:225–6).

Investigations of journalists tend to focus on their claims that journalism is a profession, and consider issues such as control over the content and structure of work, and the forms of workplace organisation created by journalists (Bagdikian 1995; Emery, Emery with Roberts 1996; Merrill 1995; Noon 1993; Reed 1991; Smith 1980). Recent technological developments have meant that the relationship between journalists and print workers is increasingly crucial, as there is now scope for journalists to undertake tasks that were previously performed by printers.

In addition to analyses of the influence of technology on print workers and journalists, a number of studies have considered the role of management and employers. These analyses indicate that proprietors and management historically conceded control of the workplace to unions, but that with the development of computer technology and its adoption in the newspaper industry, an opportunity was provided for employers to regain that control. Ozaki states, for example, that:

> a number of craft skills . . . and other traditional skills . . . which used to give their holders a certain discretion in organizing their work, have mostly been eliminated by the computerisation of work, and the new skills needed for operating computerised machines are increasingly under management control. There seems to have been a transfer of the sources of skills and knowledge from workers to management (1992:3; see also Frenkel and Weakliem 1989).

A review of the research on relationships between technological innovation and workplace reorganisation in the newspaper industry indicates that there is a need to examine to what extent such associations are influenced by relations between workers – in particular, print workers, journalists and their unions – and relations among employers, management, unions and workers. The importance of these actors and their relations in understanding processes of reorganisation in the newspaper industry is also suggested by the relational model of workplace reorganisation which guides this study. This chapter identifies the key newspaper industry actors in Britain, Australia and the United States, and analyses how they have approached the issue of technological innovation. The framework will then be established to examine the relational model in an analysis of workplace reorganisation in the newspaper holdings of News Corporation.

The National Newspaper Industry in Britain

Until the mid-1980s national newspapers in Britain were published predominantly in Fleet Street, London. Popular commentary on workplace relations in Fleet Street generally portrays it as a unique industrial phenomenon, marked by conflict, ineptitude, greed and hunger for power. Simon Jenkins, a former editor of *The Times*, wrote:

> The conflicts which consumed most of British industry [in the 1970s] seized Fleet Street with a vengeance. Years of incomes policy led to the strengthening of shop-floor power structures and payments, which have in turn broken the worker's loyalty to his employer and his union leadership, leaving the latter largely unable to police its members. National newspapers have become a sick and subsidised industry in which cynicism and bloody-mindedness often seem the sole orders on the production line each day (1979:121–2).

Central to the industrial phenomenon of Fleet Street were battles over control of the workplace, often linked to disputes about the introduction of new technology. In particular, the print unions and chapels of Fleet Street were represented as having obstructed the pace of technological development; and as having exercised unacceptable dominance over management. Wintour writes, for example, of Fleet Street chapels 'exploiting their power quite ruthlessly and with no thought of tomorrow' (1989:240), and of the 'arrogance' of unions (1989:243). At the same time, management in Fleet Street was castigated for having allowed labour to gain what was considered to be excessive influence and power in the workplace (Littleton 1992; Martin 1981, Sisson 1975).

Relations within the newspaper industry, in particular relating to ownership, had also attracted the attention of the government. From 1947 to the late 1970s there were three Royal Commissions into the press, focusing primarily on trends toward monopoly ownership in the industry. In spite of the concern about potential monopoly ownership of the press in the postwar period, it has been argued that such trends still exist. One report recently claimed that:

> the British national newspaper industry in the late 1980s and early 1990s is still relatively concentrated having been dominated by two major proprietors, Mr Rupert Murdoch and the late Mr Robert Maxwell, who controlled 36 per cent and 27 per cent of the national market respectively (AHRSC 1993:23).

Actors in the British Newspaper Industry

Workers in the British newspaper industry have been organised into various combinations of unions since at least the 1680s, and print workers are believed to have organised the first authentic trade unions in Britain (Littleton 1992). The number of trade unions representing workers in the industry has, however, varied considerably. In 1948 there were ten craft unions in the print industry, for example, but by the early 1980s the National Graphical Association (NGA) was the single craft union for print workers across British industry. On the non-craft manual side, two pre-existing unions amalgamated in 1982 to form the Society of Graphical and Allied Trades '82 (SOGAT), while journalists were represented primarily by the National Union of Journalists (NUJ). In the mid-1980s, however, talks about further amalgamations occurred, and in 1991 the Graphical, Paper and Media Union resulted from a merger between the NGA and SOGAT. In 1998, the membership of the GPMU was over 200,000, while membership of the NUJ was more than 25,000 (GPMU 1998; NUJ 1998). The Electrical, Electronic, Telecommunications and Plumbing Union (EETPU), representing electrical and other technical workers, was to become increasingly influential in the newspaper industry in the 1980s as the use of computer technology became more widespread.

Until the latter part of the 1980s, employers were represented primarily by the Newspaper Publishers' Association (NPA), which until 1968 had operated under the name Newspaper Proprietors' Association (NPA) (Gennard 1990). The Association represented employers in negotiations with trade unions over issues such as wages and conditions of work, and 'negotiated industry-wide rates and standards' (Littleton 1992:11). With a general decline in multi-employer bargaining in British industry as a whole in the 1980s, and with an increased focus on local workplace bargaining, the Newspaper Publishers' Association notified the print unions that there would be no national negotiations in 1987. Later that year the Council of the Newspaper Publishers' Association announced the disbanding of its industrial relations department. Rather than bargaining through an employers' association, companies began to bargain with unions at an enterprise level (Gennard 1990; Oram 1987).

Compared with other industries in Britain, the unions and their local workplace branches, or chapels, in the newspaper industry were successful in gaining considerable influence over workplace practices and relations. The chapels were often able to operate independently of the national union leadership, and were responsible for recruitment of employees from the relevant local branch of the union, staffing arrangements, holiday shifts and overtime rates, and certain promotional decisions. In addition, all relationships between the employer and the employee, except for purely personal issues, were dealt with by chapel officials. Within the production process itself, there were strict union demarcations between jobs, with responsibilities for tasks divided between unions. According to management, the unions and chapels successfully established a number of so-called restrictive practices, including a 'blow' or work-break system, overstaffing, restrictions on output, payment for overtime not worked, and restrictions on flexibility and mobility based in demarcation arrangements (Willman 1986).

Factors which contributed to this particular development of workplace relations in the newspaper industry included the perishable nature of newspapers as a product, reliance on a highly skilled craft workforce, the existence of total unionisation in the workforce, and a closed shop. The craft unions, including in particular the NGA, also 'maintained tight control over the production process' (Gennard and Dunn 1983:17). Craft unions achieved this control through three primary means. First, the union limited the number of apprenticeships; second, access to craft jobs was made dependent on having served an apprenticeship; and third, unions required that members not work for employers who used non-union labour (Gennard and Dunn 1983). On the non-craft manual side, SOGAT also exercised some control in the workplace through the operation of a closed shop.

The National Graphical Association (NGA)

Representing skilled craft workers, the National Graphical Association (NGA) has been intimately involved in technological developments in the newspaper industry. By 1985, it was a signatory to a number of new technology agreements (NTAs) (McLoughlin and Clark 1994; NGA 1986b), and recognised that while technological developments could not be resisted completely, they represented a fundamental challenge to the future of its members. In 1983, Tony Dubbins, the General Secretary of the NGA, wrote in the NGA magazine *Print* that '[t]he issue of technological change is one of the most fundamental facing us in Britain today' (1983:8). He argued that the introduction of technology into the newspaper industry should not be used by employers simply as a means of reducing the labour force, removing trade unions as a presence, and restoring unfettered managerial control in the workplace. In relation to the NGA specifically, the potential for the loss of use of skill related to technological innovation was great, with Dubbins commenting that '[i]f technological change means that a craft is deskilled, and the individual finds that it is no longer essential for his contribution as a craftsman to produce a print end product, his whole social standing and status is affected' (1983:8). To prevent such technologically related de-skilling occurring, Dubbins proposed that employers had a social obligation to consult with unions, and to provide guarantees concerning wages. Also there were to be no compulsory redundancies, and health and safety conditions for those employed in the industry should be monitored. Such views were also represented in official NGA documents, including a document of the NGA National Council endorsed at the 1984 Biennial Delegate Conference entitled *The Way Forward – New Technology in the Provincial Newspaper Industry* (NGA 1984). In this document, the NGA stated that it was prepared to accept technological developments provided that the union and its members would be enriched, both in terms of organisational capacity and work environment. Attempts by proprietors to introduce technology into the workplace without accepting the needs of both workers and the union would be opposed.

This concern with the social obligations of employers to their employees was counter to the neo-liberal individualistic ideology of the Conservative Party government which placed the responsibility for employment with the individual employer and worker in a free market system. Despite the desires of the NGA to protect its members' jobs, the market reform policies enacted by the Thatcher government in conjunction with technological developments were to weaken its bargaining position.

The Society of Graphical and Allied Trades (SOGAT)

Representing the non-craft white collar workers in the industry, the union leadership of SOGAT was aware of the challenge of technological development to unions in the print and newspaper industry. In 1985, SOGAT sent a study group of high level union officials, including the general secretary, to the United States and Canada to observe the influence of technological developments on the newspaper industry. The fundamental conclusion of the report they subsequently published was that simply opposing the introduction of different technology into the workplace was not an option for print unions, and would be a rapid road to de-unionisation. After visiting newspapers including the *Wall Street Journal, USA Today* and the Toronto *Globe and Mail,* the group concluded that union solidarity and cooperation were needed in the attempt to ensure that employees benefited from new technology; that employers in the United States had used technological change to cut jobs, end skill demarcations, and make large scale economies; and that developments in technology would affect not only production, but also the distribution of newspapers. In regard to SOGAT members, computer-based direct input and other information technologies threatened all jobs, with even newspaper libraries being 'ripe for computerisation' (SOGAT 1986:11). As a result, SOGAT policy was to recognise that technological investment would occur, and that it was the role of the union to attempt to ensure that its members at least shared in the benefits of any such investment.

In July 1985, Brenda Dean, the General Secretary of SOGAT, stated:

> Now new technology is being introduced into the provincial newspapers, and no doubt in a big way in the near future in national newspapers.
>
> The ultimate step forward in these areas is, of course, the introduction of direct input, which allows copy input from visual display units to be fed straight into the computer and ultimately out again for final printing. Like other forms of new technology this process is no respecter of union membership cards.
>
> Direct input systems cross the boundaries of trade union organisation by ourselves, the National Union of Journalists and the NGA.
>
> In some instances it simply combines jobs previously done by the NUJ and the NGA. In others, jobs previously done by SOGAT and the NGA will no longer be necessary. And in certain instances it combines jobs previously done by members of all three unions – or members of no union!
>
> This technology is posing severe problems for all three unions in the printing industry and is undoubtedly a severe threat to jobs . . . *The traumatic effects of a combination of destructive, negative Government policies together with new technology is only too apparent.*
>
> That combination has cost this union over 40,000 jobs in the last five years (Dean 1985:2–3, emphasis added).

As with the NGA, SOGAT was concerned about the possible influence of technological developments on its organisational basis, but at the same time recognised that simple opposition was not sufficient.

The National Union of Journalists (NUJ)

As a union organisation, historically the NUJ has not been as influential as the other unions involved in the newspaper industry. An important source of this relative lack of influence is the diverse range of occupations captured by the term 'journalist' (Noon 1993). It is a generic term covering 'not only newspaper reporters, sub-editors and photographers, but also radio and TV editorial staff, public relations officers, press officers, magazine staff, editorial staff in book publishing, freelances, press agency staff, creative artists, and teletext and viewdata staff' (Noon 1993:107). With such a diverse range of members, with varying and often different interests, it has often been difficult for the union to organise concerted action. In addition, many journalists consider themselves to be skilled professionals, having more in common with management than with shopfloor workers. The consequence is that historically it has been difficult for the NUJ to sustain concerted action in the same manner as SOGAT or the NGA.

While much of the concern of the NUJ in relation to computer technology has focused on health and safety issues for journalists, including repetitive strain injury and eye conditions, technological developments within the industry also have the potential to increase the influence of the NUJ as an organisation. This is principally because many tasks formerly undertaken by printers can now be performed by journalists, primarily through electronic pagination, giving journalists a much more central role in the production process. With the role of journalists changing in the industry, the NUJ has been successful in a number of workplaces in negotiating agreements for its members with employers over issues such as job design, ergonomics, health and safety, training and transfer, employment levels, and new technology payments (NUJ 1977; Noon 1993).

The Electrical, Electronic, Telecommunications, and Plumbing Union (EETPU)

Developments in the 1980s resulted in the EETPU becoming increasingly visible and important in the newspaper industry. Representing electrical and other technical workers, one strategy adopted by the EETPU to expand its membership base was to move into already unionised industries and sign single-union, no-strike deals with employers who wanted to employ a new workforce. Technological developments in industries such as the newspaper industry were crucial to this strategy, particularly to the extent that such developments meant that the computer-literate technician members of the EETPU could do jobs that previously required craft skills. The willingness of the EETPU to sign single-union deals provided employers with a means of reorganising their employees technologically and replacing their existing workforce if it was opposed to reorganisation. The position of the EETPU was presented emphatically by its General Secretary, Eric Hammond, in an EETPU publication, *Training for a Secure Future: Towards 2000.* He proclaimed:

> The EETPU is different and proud of it. We are unequivocal. Technological progress is vital to industrial survival. Our concern is to ensure that it is successfully harnessed, not fearfully rejected by the industrial backwoodsmen in some shortsighted emotional spasm. We seek an updated and dynamic economy in which innovation and investment is an urgent priority.
> We cooperate readily with fair employers. They and our members alike gain immensely from the enlightened self-interest involved in doing so and from the presence of our genuinely representative, independent and trustworthy organisation at their workplace. Many managements have good cause to be grateful for our efficiency, expertise and effort.
> Our members are ready for the challenge ahead. They look especially to the development of the 'sunrise' and service industries where so many of

them are already employed. They are aware that the nation's industrial future and our own are permanently and inextricably linked with the critical fortunes of industrial change (n.d.:2).

While this approach allowed the EETPU to enter into agreements with various companies in a variety of industries in the 1980s, it also provoked hostility from, and confrontation with, the union movement. In 1988 the EETPU was suspended and subsequently expelled from the TUC for breaching the Trades Union Congress Bridlington Principles, whereby unions are forbidden from either poaching the members of another union or from making agreements which ignore the legitimate interests of other unions (TUC 1988). In 1992 the EETPU amalgamated with the TUC-affiliated Amalgamated Engineering Union to form the Amalgamated Engineering and Electrical Union, which in 1995 had a membership of 835,000 (Clement 1992; *New Statesman and Society* 1995).

The relationships among these various unions, in a period of institutional and social transformation in the 1980s, were to have a dramatic influence on the outcome of the transformation processes undertaken by News International. In particular, the company was able to exploit the divisions that arose within the labour movement to ensure that it was able to hire a workforce to ensure that production continued at its new site.

Employers and Management

Contributing to the powerful influence exerted by print unions in the newspaper industry prior to the 1980s were the idiosyncratic workplace and commercial practices of employers and management. According to Simon Jenkins (1979), a former editor of *The Times*, the upheavals confronting newspapers in the late 1970s, and the continuing conflict between managers and shopfloor workers, were due primarily to the role of the proprietors in the industry. This concern, extended to management, was also recorded by various Royal Commissions and government reports which identified lack of managerial firmness and a lack of employer solidarity for the continual disputes in Fleet Street and for problems of profitability (Cockburn 1983).

Despite the antagonism that typified relations in Fleet Street, there had been attempts by employers to introduce technological innovation through negotiation at various times. For example, in 1976 The Joint Standing Committee for National Newspapers (JSC) launched a *Programme for Action* (JSC 1976), while in 1983 the Newspaper Society announced Project Breakthrough. These attempts to reorganise the newspaper industry through the introduction of technology were ultimately defeated in the face of union opposition based on the threat of massive job loss that such proposed changes would entail. Technological developments in combination with a transformed institutional and societal context were, however, to have a major influence on the capacity of management to introduce workplace reorganisation, giving them a powerful bargaining tool in negotiations.

New Technology and Workplace Reorganisation in the British Newspaper Industry: Experiences in the 1970s and 1980s

In the 1970s and early 1980s, proprietors in the British newspaper industry attempted to introduce technology on an individual or company basis on a number of occasions, often resulting in dispute. The common feature of many of these disputes over technology was that 'the employer was attempting to remove the NGA from all or parts of the company' (Gennard 1990:468). A key dispute involving national newspapers occurred in 1978–79, and involved Times Newspapers Limited, then owned by the Canadian, Roy Thomson. In 1976 the company had purchased high-technology composing room equipment valued at £3 million without consulting either the unions or chapels at the newspaper. The ensuing dispute centred around company insistence that 'the print unions negotiate against arbitrarily imposed deadlines and that the NGA surrender control of the original keystroke', and lasted for twelve months (Gennard 1990:479). In 1977, the company produced a document entitled *Opportunity for Success*, in which it laid out the necessity for single keystroking. Single keystroking meant the introduction of a direct entry system whereby journalists, for example, could enter stories directly onto computers. This would mean the elimination of the 'second keystroke', the method through which typesetters and compositors had retyped the original copy of the journalists into the system (Gennard and Dunn 1983; Noon 1993). The strategy was rejected by the NGA composing chapels at the newspaper, whose members were threatened with job loss by the plan. In 1978, management delivered an ultimatum demanding that the unions 'negotiate agreements committing them to continuous, uninterrupted production, with punitive sanctions to deter unofficial strikes' (Cockburn 1983:81). When agreement was not reached, a nine-month lock-out followed in which the NGA refused to give in to management demands over keystroking. When the dispute was ultimately settled, little had been decided about the issue of technology, the company had lost substantial profits and conceded to wage increases. The ultimate result was that Thomson put the paper up for sale.

While the *Times* dispute typified industrial relations in the national newspaper industry until the mid-1980s, developments in the early 1980s in the north of England were to send a warning to all involved in the industry that major upheavals were ahead. In July 1983, the owner of the Messenger Newspaper Group, Eddy Shah, dismissed NGA members who had withdrawn their labour because of conflict over the terms and conditions to be applied in new typesetting and printing operations. Shah had also advertised positions available for non-union labour in posts held by NGA members, and refused to recognise the NGA closed shop. There followed extensive picketing of the plant in Stockport, and at other Messenger plants, as well as 'blacking' of Messenger Group activities. When the NGA boycotted labour injunction proceedings, in which the company sought to prevent such actions, Shah succeeded in obtaining sanctions against the union for contempt of court. After the union refused to pay court imposed fines, its assets were sequestrated, while production and distribution of the newspaper continued (Elgar and Simpson 1993; Littleton 1992). As Elgar and Simpson suggested, the lessons from this dispute were that 'existing labour practices in provincial newspapers could be changed and that industrial action to [prevent] this might be successfully overcome with the aid of the law' (1993:92).

At the same time, Shah began plans to launch a national daily newspaper to be called *Today*. Spending expansively on advanced production technology, '*Today* planned to by-pass traditional composing methods by placing responsibility for typesetting and page layout in editorial offices through direct input by journalists and staff trained to operate electronic pagination systems' (Littleton 1992:36). Shah also bought a truck fleet to allow for the distribution of the newspaper independently of the railway system. Two further factors were crucial. First, the entire operation would occur outside of Fleet Street, and second, Shah obtained a labour force for the newspaper by negotiating a single-union deal with the EETPU which enabled him to avoid any dealings with the print unions. In conjunction with disputes at the *Kent Messenger* and the *Wolverhampton Express*, these activities resulted in new work practices and new technologies being introduced into the newspaper workplace, despite the opposition of the print unions (Elgar and Simpson 1993; Gennard 1990).

These were significant developments in, the newspaper industry, signalling a willingness on the part of a proprietor to use the law and the changing social and institutional context to confront and succeed against the unions. Print unions in the British newspaper industry had historically been very powerful actors in the newspaper industry. The combination of new technology and a Conservative Party government intent on transforming the institutional framework of labour relations in Britain, however, provided a new context for production of newspapers. When combined with the rise of entrepreneurial owners such as Eddie Shah, and divisions within the union movement, the circumstances were set for a major reorganisation of relations in the industry. This context was to be seized upon in the strategy developed by Murdoch and News International in the shift from Fleet Street to Wapping. At the same time, actors in the newspaper industries in Australia and the United Sates were facing similar challenges.

The Newspaper Industry in Australia

The Australian Printing and Kindred Industries Union (PKIU) observed in 1980 that within the printing industry:

> changes are taking place with such rapidity that as soon as one technique is studied and mastered by the Craftsmen, it has to be thrown out and replaced by something newer. The tragic fact that emerges is that each new step towards automation reduces the cost of labour, because each step is labour saving. It would seem that zero engagement is the first step and zero staff the next (Bennett 1980b:1).

As in Britain, workers in the Australian newspaper industry in the 1980s were confronted with the challenge of computer technology which could both automate production techniques previously performed by human labour, and 'informate production' so that three-dimensional production processes could be transferred to, and undertaken on, a video screen (Evatt Foundation 1995:143). While these developments threatened the position of craft workers in the industry, they also had the potential to expand the tasks performed by white collar workers, such as journalists. As the 1980s progressed, these technological developments were accompanied by a fundamental reorganisation of the workplace in the Australian newspaper industry.

An Australian House of Representatives Select Committee (AHRSC) on the Print Media reported in 1992 that '[t]he Australian print media is highly concentrated. In almost every sector of the industry, one or two groups dominate in terms of the number of publications and the related circulation under their control' (1992:xvii). In the mid-1980s, the major news print media groups in metropolitan Australia were News Corporation Limited, John Fairfax and Sons Limited, and the Herald and Weekly Times Limited (AHRSC 1992). With the takeover of the Herald and Weekly Times by News Limited in 1987, ownership of metropolitan newspapers was effectively reduced to two major groups, News Limited and Fairfax, with News Limited newspapers comprising 62.8 per cent and Fairfax controlling 20.4 per cent of the total circulation of metropolitan dailies and Sunday newspapers (AHRSC 1992:102).

Actors in the Australian Newspaper Industry

While many unions have been involved in the newspaper industry in Australia, two of those most affected by technological innovation have been the Printing and Kindred Industries Union and the Australian Journalists' Association. Print workers in Australia have been organised into unions since 1857, and from 1966 to January 1995 were represented nationally by the Printing and Kindred Industries Union (PKIU). The PKIU itself had been formed as a result of an amalgamation in 1966 between the Printing Industries Employees' Union of Australia and the Australian Printing Trades Employees' Union, and at times represented nearly 90 per cent of union members in the printing industry (Frenkel and Weakliem 1989). In 1966, membership of the PKIU was around 50,000 and rose to a peak of 60,000 in 1970. There was then a steady decline in membership, associated with cost economies and related technological innovation requiring less PKIU labour, resulting in the PKIU having a membership of 43,000 in 1992 (Cahill 1993).

Membership of the PKIU was made up of 'clerical, administrative, and production employees employed in the Printing and Kindred Industries', with some '50 percent of members being tradespersons with the remainder ranging from semi-skilled to professionally qualified employees' (Cahill 1993:121). As in Britain, the national PKIU was represented at the local workplace by chapels, each of which 'is a relatively autonomous unit within the Branch structure of the Union ... [which] ... operates to a considerable extent as an employee committee within the Company' (MacIntosh 1984:35). While discussion of the organisation of print and production workers in this study will focus on the PKIU, in 1995 the PKIU amalgamated with the Automotive, Food, Metals, and Engineering Union (AFMEU) to become part of the latter union organisation (*Metal Worker* 1994), and by 1998 was a part of the Australian Manufacturing Workers Union (AMWU) – a union with 200,000 members (AMWU 1998).

Until 1992, journalists in the Australian newspaper industry were represented by the federally registered Australian Journalists' Association (AJA) which was also affected by technological innovation in the 1980s. As well as acting in the industrial sphere on matters including wages, health and safety, and job grading, the AJA sought to promote membership awareness of professionalism in journalism, including professional recognition, education, and ethics. A federal president of the AJA stated, 'The AJA is both a professional association and a trade union. This necessarily involves some difficulties but it does not alter the fact that the two roles are, and should be, complementary' (Lloyd 1985:288). One consequence of this dual role was that the AJA was not part of the ACTU. Members considered themselves to be white collar professional workers distinct from the majority of blue collar union affiliates of the ACTU. With 12,000 members in the mid-1980s, the AJA was the largest white collar union not associated with the ACTU (Lloyd 1985). On 18 May 1992 the AJA amalgamated with Actors Equity of Australia and the Australian Theatrical and Amusement Employees Association to form the Media, Entertainment and Arts Alliance (MEAA). This is a federal union somewhat ironically affiliated with the ACTU, which by 1998 had a membership of 36,000 (*Alliance* 1993b; MEAA 1998).

Employers have had a number of different organisations including Country Press Australia, Regional Dailies of Australia, and the Printing and Allied Trades Employers' Federation of Australia (PATEFA), which is a member of the Confederation of Australian Industry (CAI). Such associations support proprietors of newspapers in various ways, from representing them in award hearings in the Industrial Relations Commission to providing information about new developments in the industry. PATEFA was active in the industrial sphere in the 1980s, disseminating information about new technology, opposing the claims of the PKIU for a thirty-five-hour week, and arguing against the ALP–ACTU Accord (*Graphix* 1981a, 1983a, 1984b, 1992, 1993b, 1993c). Other organisations, including the Pacific Area Newspaper Publishers' Association (PANPA) which considered its main aim 'to spread knowledge of the new technology among publishing executives' (*Graphix* 1983b:2), were among the associations to which employers and executives in the newspaper industry turned for assistance over matters including technological innovation, and negotiations with unions, employees, and government institutions.

The Printing and Kindred Industries Union (PKIU)

The PKIU was active in supporting and promoting a policy of awareness and action in regard to technology and its consequences for the print workforce and workplace relations. While acknowledging that a gradual introduction of new technologies into the industry occurred in the 1970s, the beginning of the 1980s was considered to be the threshold for significant technological developments, with the industry facing 'some of the greatest challenges since the introduction of movable type, the application of steam power to the printing press or, indeed, the invention of the wheel itself' (Bennett 1980a:1). It was claimed that 'the work of the industry is not static. Change has become a natural phenomenon. There have been, and will be, constant changes in the skill requirements whereby continual training and re-training will be necessary on an ever-increasing scale' (Bennett 1980a:1). In anticipation of such developments, it was proposed that '[t]he union does not oppose New Technology. It has decided that as it is inevitable then the Union's function is to use it as a means of improving the wages and conditions of members so that they will share in the benefits of improved productivity' (Bennett 1980a:1). The call was for engagement but struggle where necessary, arguing that while the union would enter into negotiation over technology, members would need to be prepared to take industrial action in certain circumstances.

In 1983 the PKIU federal council established a New Technology Committee with the aim of 'obtaining not only an on-going descriptive summary of technological developments in areas of our concern but to investigate and set guidelines in order that members may benefit from such developments' (Cahill 1987:1). The formation of the committee was a new development in union policy, with the focus shifting from providing information only on the form of technology being introduced into the workplace to a focus on the potential effect of technology on people. Cahill, as federal secretary of the PKIU, emphasised that the union did not oppose technology aimed at enhancing the industry's value and contribution to the economy, but was concerned about technology being introduced by companies 'to destroy proprietary rights of labour or its say in our industry's management much to the detriment of the industry and Australia's economy' (Cahill 1987:1). By arguing for proprietary rights and for a voice in decision-making, the PKIU was showing that it expected to have a role in the introduction of technology into the workplace.

As part of this process of engagement in debate over technology, the PKIU developed a Model Technology Agreement, which contained their basic requirements for any negotiated agreement in relation to new technology and for the regulation of its influence on the workplace and on workplace relations. The Agreement was proposed on the basis that the introduction of new technology need not necessarily lead to job loss, workforce de-skilling or reduced job satisfaction, but that such outcomes would not be avoided unless there was regulated and joint negotiation between the workplace parties over how technology was to be introduced, and how its effects were to be managed. It was stated that there was a need to consider not only the economic effects of technology, but the social effects on employees, including job content and skills (PKIU 1987). As a corollary to the Agreement, the PKIU urged its members to be active in negotiations concerning new technology because union members were 'the lynch pin in any information sharing exercise' (PKIU 1987:49).

The policy of the PKIU in the 1980s conformed with that of both the ACTU and the federal Labor government in supporting negotiation and consultation over the introduction of technology into the workplace. Contrary to the experiences of print workers in Britain under the Conservative Party government, print workers in Australia in the 1980s and into the 1990s operated in an institutional context supportive of negotiation and consultation. In pursuing this goal, relations between the PKIU and the Australian Journalists' Association were vital.

The Australian Journalists' Association (AJA)

The attitude of the AJA towards the introduction of new technology in the 1980s was one of optimism, and an acceptance of the necessity and inevitability of the developments. While the AJA expressed doubts about the effectiveness of video display terminals (VDTs) in the production process, union officials believed that eventually the cost reductions that would accompany this innovation would ensure the economic survival of the newspaper industry in the face of competition from developments in electronics which assisted the growth of alternative media forms (Lloyd 1985). The AJA did not simply accept the introduction of technology into the workplace, however, but sought to influence the manner in which it was introduced, and its subsequent impact on workplace relations, including crucial issues of health and safety. At the 1983 AJA Annual Conference, for example, the AJA Federal Council ruled that 'before the AJA accepts the expansion of existing technologies or the introduction of new technology the union consults at federal level in conjunction with branches, with relevant other unions in the industry and reports the results to Federal Executive and branch committees' (*Graphix* 1984a:1). Consultation and exchange of knowledge were considered key elements of AJA strategy.

The leadership of the AJA also claimed that journalists should receive a share of extra profits made by newspapers with the introduction of VDTs. According to Lawrence, the President of the AJA, 'The essential argument justifying extra money for all remaining employees [after redundancies], be they journalists or printers, is that they deserved a share of the benefits of new technology' (1983a:5). Lawrence indicated that he believed 'sensible employers' would recognise 'the moral obligation to share the benefits of new technology with employees and reach agreement in negotiation' (1983a:5). As with the PKIU, the AJA was determined to be involved in negotiations over the introduction of new technology into the workplace. Job loss, modifications of job description and function and health issues were all of concern to the AJA in its negotiations with management and owners regarding the relation between technology and workplace reorganisation.

Central to union strategy about new technology were the relations between the AJA and the PKIU over workplace reorganisation. Divisions between the unions had developed in a strike over technological innovation at Fairfax newspapers in the 1970s, a split referred to by the President of the AJA as 'a tragedy – a tragedy caused as much as anything else by the lack of effective communication between the two organisations about their real fears and problems' (Lawrence 1983a:5). Evidence of the lack of communication in the 1970s is reflected in the observation that '[r]eporters and sub editors saw two things in it [that is, technological innovation] for them ... money and power ... Job sharing with printers who would be wiped out was not an issue at the time' (Lloyd 1985:276). Relations between the unions did improve, however, and they entered into agreements over the introduction and use of new technology in the workplace (*Journalist* 1986, 1988). Through the 1980s, agreements between the PKIU and the AJA concerning new technology became an important part of the struggle to control the manner in which new technology was introduced into, and influenced, the workplace and workplace relations.

Management and Employers

For management and employers in the newspaper industry, the introduction of technology into the workplace was considered to be essential if individual newspapers were to improve the quality of their product and remain competitive. In his 1993 'Chief Executive's Report', for example, Rupert Murdoch stated that the future of the media was with companies that could develop a vision for innovation in the face of the ongoing technological revolution, and could then transform that vision into reality through dynamic leadership. He went on to say that while News Corporation had a proud history in media industries including newspapers, the company was constantly exploring future possibilities both for taking advantage of the opportunities offered by the globalisation of the media, and for responding to the changing demands of readers (1993).

Management and employers also left little doubt that technological innovation required fundamental reorganisation of the workplace. Larry Lamb, then an editor of the *Australian* newspaper owned by News

Limited and a deputy chair of the company, addressed the Advertising
Club Ltd in Sydney in 1982:

> It has become fashionable in those areas of my industry where editorial
> initiative, energy and expertise seem to have dried up, for proprietors to pin
> all their hopes on the so called 'new' technology . . . We should welcome the
> new technology, of course, but there is one thing we should always remember
> about it. It does not make better newspapers. It does not even make faster
> newspapers. I do not know a newspaper in the world which has been
> improved by it . . . The heart of a newspaper lies not in machinery, however
> sophisticated, but in people. The new technology is useful only if it enables
> an employer dramatically to reduce his labour costs. In News Limited we have
> achieved exactly that. Those who buy it by paying even higher wages than
> before to people whose jobs have become so simple that they could be done
> by any competent school leaver are buying dust and ashes (1982:2).

The link drawn by management and employers between introducing
technology and the need for workplace reorganisation is revealed in
such statements. Management also exhorted workers to cooperate in
workplace reorganisation. In 1988, the former industrial relations
officer of PATEFA argued that there was a need for employees to
communicate with management, and to play a participatory role in
business. He suggested that a unity of outlook among employers
and employees was required, possibly involving the creation of single
organisations representing unions and employers (*Graphix* 1988a).
Management proposed that it was in the common interest of all
involved in the industry to pursue technological innovation.

New Technology and Workplace Reorganisation in the Australian Newspaper Industry: Experiences in the 1970s and 1980s

Unions, employers and their representatives, governments and state
institutions, including the Industrial Relations Commission, have all
been key actors in affecting technologically-related workplace reorganisation in the Australian newspaper industry. In 1980 a study entitled
Technological Change in Australia, but referred to as the *Myers Report*
in honour of its Chair, Professor Rupert H. Myers, was published by
the Committee of Inquiry into Technological Change in Australia.
Commenting on the activities of Australian newspaper firms owned by
John Fairfax and Sons, the Herald and Weekly Times, and David Syme
and Company Limited, the report indicated that '[o]ver a short period
of time [in the late 1970s] the industry has become a leading innovator
in the adoption of computer-based technology' (*Myers Report* 1980:147).
From an analysis of negotiations about technology and workplace

reorganisation, the report revealed that the major reasons for reorganisation, which included reducing the labour content in production processes, speeding up production, and reducing errors, were all 'associated with the attempt to lower costs' (*Myers Report* 1980:161). In the case of Fairfax, where the attempted implementation of technology-related workplace reorganisation was accompanied by a sixty-day strike, no consultation was entered into with unions and no information was provided to employees 'until after the company had announced its automation plans, and had signed a contract for the delivery of the system' (*Myers Report* 1980:165). At both David Syme and Company Limited and at the Herald and Weekly Times, meetings were held with PKIU representatives to outline plans for technological innovation more than two years before either company signed a contract for automated systems, even though no commitment had been made by the companies to any one particular computer system (*Myers Report* 1980). Indeed, the latter two companies joined together for discussions with the PKIU. There were also differences in the terms of the benefits that employees obtained from the introduction of technology. At Fairfax, an estimated 432 members of the PKIU, which held 659 positions in the workplace, took compulsory redundancies. In contrast, at the other two firms a settlement was reached that there would be no compulsory redundancies with management and the unions agreed that work hours were to be reduced, voluntary retirement packages would be offered and training programs would be established (*Myers Report* 1980). In seeking to explain the contrasting approaches of the companies to technological innovation, the *Myers Report* suggested that both David Syme and Company Limited and the Herald and Weekly Times had the benefit of introducing technology after Fairfax, and were aware that lack of early negotiation with unions would create the possibility of a financially devastating strike.

Another important event in the dispute at Fairfax was the breakdown of relations between production workers, journalists and clerical workers over job demarcations in the reorganised workplace. While PKIU production workers sought to maintain control over the labour process to the exclusion of other workers, the AJA, in agreement with the company, argued that journalists should have the right to operate VDTs, input their own copy into computer banks, subedit copy, and issue instructions to the computer for typesetting. The Federated Clerks' Union (FCU) was also involved and argued, with the support of the Fairfax company, for the right to operate VDTs in relation to advertising. These arguments over demarcation were made before Justice Cahill in the New South Wales Industrial Relations Commission, and he ruled in favour of the company, and the AJA and FCU. He decided that journalists had the right to use new technology to create and edit material, and that clerical workers could use VDTs in relation to classified advertisements. While PKIU members were granted exclusive jurisdiction in some areas, control over production work was narrowed considerably by this decision (Lloyd 1985; *Myers Report* 1980). In his finding, Justice Cahill stated:

> Under the new [computer] system, the typesetting function, in the sense of a separate function as undertaken under the existing system by PKIU members, has disappeared, a situation which has come about as the result of the journalists carrying out their legitimate primary function of story creation and recording (Cahill decision 1977:537).

According to the authors of the *Myers Report*, this decision:

> set an important precedent in the industry by concluding that the typesetting function as it existed in the past had disappeared as a result of the adoption of new computer-based technology. Journalists and clerical assistants using video display terminals were able to send typesetting instructions into the system, thus eliminating the need for compositors in the traditional sense. Also, since the journalists and clerical workers were expected to perform the task of proof-reading, they reduced the need for specialist proof-readers (*Myers Report* 1980:174).

While the decision in this case was specific to Fairfax, it soon became a precedent for other companies in the industry, even where reorganisation related to technology occurred without conflict (*Myers Report* 1980). Reed reported, for example, that the process of introducing computerised photocomposition into the *Age* in Melbourne in the early 1980s was based on negotiations founded on the Cahill decision. In the case of the *Age*, by making use of the demarcations specified in the Cahill decision, the agreement between management and the PKIU was 'a "defensive" agreement to facilitate technological change, although on terms which limit management's power to rationalise labour and exclude union influence' (Reed 1988:49). Reed observed further that agreements across the newspaper industry in Australia had revolved around the issue of 'the capture of the first keystroke or direct entry of copy to the computer' (1988:48). Rather than opposition to technology, struggle centred around the division of labour within the workplace in the face of reorganisation related to technological innovation.

With the election of the Australian Labor Party to federal government in the early 1980s and the establishment of the Accord between the new government and the union movement, an institutional context was created in which consultation and bargaining were proposed as appropriate means to bring about workplace reorganisation. While Australian unions were able to maintain their position as key actors in the newspaper industry, at least in part due to this more supportive institutional context, the experience of newspaper unions in the United States in the same period was to be very different.

The Newspaper Industry in the United States

Newspapers in the United States have constantly been at the forefront of technological innovation and implementation in the workplace, and newspaper companies from countries such as Australia and Britain have frequently looked to the US both for examples of technology use and for possible models of workplace organisation.

In the 1980s, following general trends in US business practice, ownership of newspapers in the US increasingly became a group enterprise. In 1960, 109 groups owned 560 newspapers, representing 30 per cent of all newspapers and 46 per cent of circulation. By 1990, 135 newspaper groups owned 1228 newspapers, while there were 383 independent dailies. The groups owned 75 per cent of all newspapers, and controlled 81 per cent of newspaper circulation (Emery, Emery with Roberts 1996). By the late 1980s many of the largest newspaper-owning companies also had extensive interests in other sources of media, including magazines, radio and television broadcasting, and cable, in addition to other non-media interests. In 1998, the largest US newspaper companies in terms of ownership were Gannett Co. Inc., Knight-Ridder, Inc., and Times Mirror Co., while the *Wall Street Journal* had the largest circulation.

The organised workforce in the US newspaper industry has historically been arranged into craft and non-craft unions. However, the unionised workers have also been fragmented and in the late 1950s, for example, there were fourteen clearly demarcated unions in the newspaper industry (Newsom 1980; Smith 1980). There have been a number of recent mergers with the International Typographical Union (ITU) and the Newspaper Guild (the Guild), for example, becoming part of the Communications Workers of America (CWA) in 1987 and 1995 respectively. In supporting the move to amalgamation, the President of the ITU wrote to members that

[the] merger will make us an integral part of, and give us access to, the power and sophistication of North America's most technologically advanced labor organisation. It is a merger that will strengthen us immediately while arming us for the labor–management conflicts and technological challenges of the remainder of the Twentieth Century . . . and beyond (McMichen 1986:17).

These former unions maintain separate sections within the CWA, while benefiting from the more extensive resources available to a larger organisation.

For the purposes of examining technological innovation and workplace reorganisation in the newspaper industry in the US, two crucial unions were the Newspaper Guild, which represented reporters and editorial workers, and the International Typographical Union, which represented many craft workers, including compositors but not the press operators. As in Australia and Britain, technology and workplace reorganisation had their most dramatic initial effect on compositors by reducing the scope of their work, and on journalists and editorial workers by potentially expanding their jurisdiction within the workplace.

A common experience across the union movement in the US has been declining membership numbers (Gifford, various editions). While the declining membership numbers and amalgamations are linked to factors identified in chapter two, including the role of the law, management hostility and economic restructuring, they are at least partially attributable to the influence of technology on the workforce employed to produce and distribute newspapers (*Guild Reporter* 1994e). By 1961, ITU membership had reached a record high level of 94,523. In 1962, however, the *Los Angeles Times* became the first US newspaper to introduce a computer into the composing room. This development allowed the company to push print craft workers to the outer edges of the typesetting process, and the number of print workers began to fall. Non-union newspapers soon followed the example of the *Los Angeles Times*, with the consequence of large reductions in the number of employed print workers. While unionised workplaces were initially able to resist attempts to reduce workplace numbers, by the end of the decade ITU membership had fallen from 94,523 to 91,848 (Eisen 1978). The introduction of VDTs into the workplace in the 1970s was associated with an acceleration of this process, with further declines in ITU and other newspaper union membership. The Newspaper Guild also experienced a decline in its membership numbers from 26,000 in 1975 to 20,000 in 1995, when it amalgamated with the CWA (Gifford, various editions). In 1998, membership of the CWA was 630,000, with 400,000 of those members in communications and related fields.

Proprietors in the newspaper industry have organised themselves around many issues, including workplace bargaining and technology innovation and investment. The American Newspaper Publishers Association (ANPA) was founded in 1887 as a trade association of daily newspapers, and by 1930 the association had 850 members. In 1951 ANPA had opened a research centre to study improvements in printing processes, and by 1960 the association had a membership of 1,200 daily newspapers. Breakthroughs in various aspects of printing techniques were made by ANPA in areas including photocomposition, off-set printing, digital computers, colour inks, and the flexographic press (Emery, Emery with Roberts 1996). In 1992, ANPA merged with other newspaper organisations to become the Newspaper Association of America, serving 1050 newspapers in the US and Canada in areas of marketing, research, technology and First Amendment issues. Although this organisation has performed many valuable functions for proprietors, as in Australia and Britain bargaining over terms and conditions of employment at the workplace with unions is performed primarily by individual proprietors and their management teams, although sometimes in conjunction with other newspaper proprietors in the same city.

Technology and Workplace Reorganisation in the US Newspaper Industry

The development of production techniques based on computer technology has had a fundamental influence on the newspaper workplace in the United States. National figures produced by the US Bureau of Census reveal that from 1970 to 1980 the percentage of newspaper employees who were typesetters and compositors declined from 12.6 per cent in 1970 to 4.1 per cent in 1980. In the same period, the number of people employed in newspapers increased from 422,657 to 510,125, but the number of typesetters and compositors fell from 50,760 to 21,128 (Cornfield 1992). Such numbers provide an indication of the sweeping reorganisation of the newspaper industry in the US from being an industry 'top-heavy' with machine operators to an industry largely staffed by white collar and clerical employees operating computers (Goltz 1989). A 1994 publication of the US Department of Labor provided further support for this claim by reporting that while 11,000 people were employed in the printing industry as precision compositors and typesetters in 1992, it was estimated that by 2005 the number would be 8,000, representing a decline in employment of 26.5 per cent. By contrast, the number of electronic pagination systems workers in the printing industry was approximately 18,000 and was predicted to rise to around 32,000 by the year 2005, an increase of nearly 80 per cent (US Department of Labor Bureau of Labor Statistics 1994). These numbers reveal that the traditional craft of composing is being replaced in the US newspaper industry by people working with electronic systems.

Developments in computer technology have also had dramatic consequences for the newsroom, where journalists and editors have increasingly assumed typesetting and proofreading functions on video display terminals. Accompanying new job titles have been created in the newsroom, including systems editor and graphics journalist. There was a 30 per cent increase in the numbers employed in the newsrooms from 43,000 in 1978 to 56,200 in 1988 (Goltz 1989). The question that arises is: What has been the approach of the relevant workplace actors to these developments?

The International Typographical Union (ITU)

Before merging with the CWA in 1987, the ITU had attempted to confront the issue of technology, primarily through cooperation with management in situations where ITU members were able to secure jurisdiction over the operation of the new machinery. There was also recognition, however, that technological developments posed a threat to the position of ITU members. In 1982, the President of the ITU, Joe Bingel, stated:

> [I]t was not until sometime in the middle 1970s that both our composing and mailrooms were introduced to what we called 'new processes innovation'. Our industry began to change and so did our union. New language was introduced in our contracts such as 'job security,' 'lifetime jobs,' 'buyouts,' etc.
>   The symbol of the new processes age of automation is the electronic computer, and with good reason. Computers can perform, in minutes and with superhuman accuracy, prodigious calculations that would take a man a lifetime to complete. Properly instructed and linked to auxiliary equipment, they can shoulder extraordinary tasks (1982:3).

While acknowledging the power of this new technology, Bingel also had a stark warning for the members of the ITU:

> As with all other powerful powers for progress of the past, the computer is, and will continue to be, disquieting to the public. Like most technological breakthroughs, automated tools and techniques, in addition to its accomplishments the computer has the immediate effect of obsoleting some skills and eliminating some jobs (1982:3).

Confronting the reality of technological innovation in the late 1960s and early 1970s, and the evident challenges to collective bargaining as a mode of conducting workplace relations, the ITU began to seek lifetime job guarantees in return for which they gave newspaper publishers flexibility in introducing and using new technology in the production centres of newspapers. The labour-saving potential of computerised technology, however, created a situation where composing rooms were soon staffed by craft workers with little to do. In this situation, publishers began to seek to 'buyout' compositors and other workers from their contracts of employment. Buyout, or termination incentive, programs represented a new period in labour–management relations in the newspaper industry in the US. They operated as financial inducements to persuade printers to give up lifetime job guarantees. Indeed, by 1981, there were 116 buyout, or termination incentive, programs between newspapers and the ITU (Newsom 1981). While union leadership sought to accommodate technological innovation, and attempted to bargain for their members, they were under no illusions as to what would be the ultimate consequence of these buyout programs. In 1978, the Secretary-General of the ITU, Thomas W. Kopeck, wrote in the *Typographical Journal*:

> Until a few years ago members of the Typographical Union understood that contract negotiations would involve hard bargaining on each side, but that both employer and employee would survive the process and life would continue. A new movement has surfaced recently indicating that some employer representatives believe they can function without us. Unbelievable as it seems, these industrial 'experts' expect us to assist in our own decimation and destruction (Kopeck 1978a:3).

Kopeck then went on to explain the consequences of these developments in more detail:

> As our traditional work was done more and more by others with our consent, employers began to complain that there was less work for our members and something must be done. About that time some genius concluded the solution was offering a sum of money to these surplus people and they would simply go away. The scheme worked well for a while until the workers began to realize that ten thousand dollars does not a capitalist make. The cost of buyouts began to climb and acceptances started to decline, forcing the employers to come up with a new device. In negotiations around the country employers are suggesting that lifetime jobs aren't really lifetime jobs at all. They argue that because our members have given up their jurisdiction for lifetime jobs, the work just won't stretch out that far, therefore the job is for the life of the contract, not the person.
>
> Once again, members who looked to the benevolent boss for security are beginning to learn the bitter lessons of their predecessors in the labor movement. Labor negotiation has always been a contest of opposing forces, but the stakes now are our very survival. Jurisdiction is the very heart of a union and cannot be measured in dollars. It has been said that automation cuts into our work drastically, which is true, but people are keying and instructing the machines and that is the function we have exchanged for the security of a lifetime job. Having accomplished that goal, the employer now suggests those not needed should be willing to go away (1978a:3).

There was clearly deep concern within the union as to the future of print workers in the newspaper industry, given the tactics being used by publishers. For some print workers, however, even if the union could be saved it was proving difficult to come to terms with what they considered to be the de-skilling influence of the technology. An article published in the *Guild Reporter* in 1980, and written by a member of the New York Typographical Union, gave full expression to the adverse influence of technology on the skilled craft worker. Edward P. Hayden, the ITU member in question, observed:

> Here is what I do for a living. I make chads. You do this by sitting at a type-setting keyboard, placing your fingers in the asdfjkl position, checking your copy and banging away. This effort produces a perforated tape that is fed to an omnivorous digital printing computer that devours it, digests it, interprets the language of holes and produces a printed page – all in a matter of minutes. It's technologically impressive and it's remarkably efficient. And I hate it.
>
> I used to be a printer, a member of the aristocracy of skilled labour . . . It was a craft, there was much to learn and learning came slowly . . . You shared a common skill, a common body of knowledge that was acquired over a period of time. It took six years or more for an apprentice to become a journeyman; I learned the necessary computer codes for typesetting in about eight weeks (1980:8).

Hayden went on to lament the lack of atmosphere in the modern production site, concluding that 'the machine has stolen my skills' (1980:8). The influence of computer technology on skill levels and on the relationship of the craft worker to the tools of the craft and to work was, and continues to be, a real concern for this category of worker in the newspaper industry.

The Newspaper Guild

For newspaper journalists and editors organised in the Newspaper Guild (the Guild), the concerns with technology as it entered the newspaper industry were at least initially of a different kind. From the 1970s to the present, much of the concern of the Guild with computer technology and video display terminals related to occupational health and safety issues. In particular, concern about the harmful effects of VDTs was heightened in the 1970s and 1980s when journalists reported repetitive strain injury, or carpal tunnel syndrome. While related, in part, to the type of machines being used, these injuries were also considered to be associated with the reorganisation of the work of journalists that occurred with the introduction of computer technology into the workplace. Particularly with the development of computer pagination technology, journalists were required not only to write stories, but to develop editing and page-setting skills that were previously the responsibility of others in the labour process. While the extra responsibilities and increasingly varied work requirements of journalists were linked to workplace health and safety issues, they have also been associated with the potential empowerment of journalists in the workplace. According to the Newspaper Guild, for example, the '[k]nowledge and training Guild members are receiving on the job is making advertising, circulation and editorial workers more difficult to replace than at any time in Guild history' (*Guild Reporter* 1994d:8). Because the tasks performed by journalists now cover so many areas of newspaper production, they are central to the labour requirements of proprietors. These developments which are related to developments in technology, however, do not insulate journalists from job loss. In their continual search for profits and cost-cutting measures, newspaper proprietors have recently been reducing the size of their reporting and editing staffs (*Nieman Reports* 1996). In particular, the growth of newspaper groups and the development of global news services gives proprietors greater capacity to reduce staffing levels among journalists by using stories and reports in more than one publication, and by gathering news information from varied sources.

In addition, the development of newspaper groups, or chains, provides employers with the resources to use an alternative labour force in the event that a newspaper is affected by a strike in relation to new technology, or some other labour issue (Bagdikian 1995). While journalists are developing broader skills, and would seem therefore to be in a strong workplace bargaining position, the development of media chains in the United States has the potential to weaken their position as those companies have more than one location from which to select their workers. Indeed, corporations with multiple investments may in many instances be able to draw on a workforce from around the world. Furthermore, given continuing patterns of unemployment and underemployment, in the event of strike activity it may be possible for companies to draw on non-union or non-sympathetic union labour from another location to ensure continued production of the newspaper.

Management and Employers

The influence of technology on workplace organisation and job demarcations suggests that a major challenge for owners and management in the newspaper industry is the process of reorganising the composition of the workforce. In some instances, management has entered into contracts with unions requiring retraining in the event of technological innovation whereby current print and production workers are trained to use new technology, and so continue their employment although in a different capacity (Bureau of National Affairs Inc., various editions). A publisher of a Louisiana newspaper reported, for example, that 'we were very fortunate we didn't have to deal with any of that [buyouts and lifetime employment guarantees]. We retained all the employees and moved them around. One of our best linotype operators ended up as an ad salesman, and he's still there' (Goltz 1989:19). This publisher went on to report that '[b]eing small made it a lot easier. When you have the right people in the workforce, they look to the future, and they really enjoy being part of the leading edge of technology' (Goltz 1989:19). By contrast, a recent edition of the *Nieman Reports* focused on cutbacks and lay-offs in newspapers (*Nieman Reports* 1996). Through a computer search of on-line newspapers, it was reported that in 1994 and 1995 there had been forty-nine separate instances of staff reductions at newspapers in the US, affecting all areas of newspapers. Some of the examples included:

> January 22, 1994. *The New York Times* announced it would reduce its non production staff, hopefully by voluntary buyouts. The paper said it was too soon to say how many employees would be affected, but Publisher Arthur Sulzberger Jr. suggested that the cuts would total 10 percent.
> March 11, 1994. Three *Boston Herald* editorial employees were laid off and 20 commercial and editorial employees accepted buyouts.
> July 18, 1994. American Publishing Company, which purchased *The Chicago Sun–Times* for $180 million on March 31, said it would lay off about 30 workers, all in the business and production departments. It had set aside $10 million to finance staff layoffs at that paper and its sister suburban publications.
> November 11, 1994. About 75 composing room workers at *The Los Angeles Daily News* have been given 60 days notice of impending layoffs. They are being replaced by a new computer pagination system.
> October 24, 1995. Knight-Ridder said that it would cut up to 250 jobs, nearly 8 percent of the workforce, at *The Philadelphia Inquirer* and *The Philadelphia Daily News* . . . The reductions will be effected by offering employees buyouts, cutting part-time hours and reducing overtime (*Nieman Reports* 1996: 17–20).

While such cut-backs are related to various factors including the desire by proprietors to increase profits and the rising costs of production associated with price increases in newsprint, they are made possible in part by technologies which allow production to occur with a smaller staff. This is particularly apparent in the case cited of the *Los Angeles Daily News* where composing room workers were replaced by a new computer pagination system. Further evidence of the processes by which technology has been introduced into the US newspaper industry is provided in the following examples of workplace reorganisation.

New Technology and Workplace Reorganisation in the US Newspaper
Industry: Early Experiences

An indication of the challenges that developments in cold-type technology would pose to workers in the newspaper industry in the US first came in 1947, when the International Typographical Union commenced strike action at newspapers in Chicago that was to last for twenty-two months as part of the struggle against the *Taft–Hartley Act 1947*. The Act was considered by the union movement to provide employers with the opportunity to hire non-union print labour, while allowing employers to discriminate against ITU members who wished to carry out traditional working agreements negotiated by their union (Smith 1980). Publication of the newspapers in Chicago continued during the strike, however, as the owners were able to use cold-type composition instead of the traditional hot metal typesetting, and thereby bypass typographers who controlled the skills required to perform the hot metal printing craft (Moghdam 1978). This strike marked a fundamental development in workplace relations in the industry. On previous occasions when print workers had taken industrial action it had not been possible for the newspaper to be produced. With the new technology this was no longer the case, and production could continue provided the publisher could find an alternative workforce. By the end of the 1950s, twenty-seven newspaper plants in the US, publishing fifty-three newspapers, had installed photocomposition machines, and thus posed a considerable threat to craft influence over the production process.

The use of computer technology in newspapers in the US was to challenge even more dramatically the bargaining position of unions in the workplace. The first newspaper strike in the US related to the use of computers occurred in November 1963, when seventy ITU printers at two Florida newspapers objected to the installation of computers into the production process and refused to work. The newspapers were able to continue production without losing a single day of publication, however, by employing an alternative production force to operate the new equipment (Moghdam 1978).

While these developments were important in setting the stage for further developments in the newspaper industry, perhaps the defining moment in discovering the influence that computer technology would have on practices of workplace negotiation and workplace reorganisation came with the 1975 strike at the *Washington Post* (the *Post*) which was to last for four months. Not only was this a major newspaper, but the events were to be used as a model of development at other newspapers, including newspapers in the union stronghold of New York City in the late 1970s. As the strike progressed, it brought vividly to the forefront issues regarding technology, and the accompanying potential for workplace reorganisation.

During the 1960s, press operators, printers and other production workers at the *Post* had succeeded in bargaining for wage increases based on factors such as a period of prosperity for the newspaper, generally favourable economic conditions, and the ability of press operators to halt production through strike activity. As bargaining rounds progressed through the late 1960s and into the 1970s the workers succeeded in winning more wage increases from the company. At the same time, the owner and the management of the newspaper were unhappy with the quality of production, and also with the continued union demands for wage increases. They also claimed that press operators were engaging in slow-down tactics in the production process which were costing the company money. Arguing that the newspaper was not sufficiently profitable, management at the *Post* began to seek means to challenge what they perceived to be the excessive influence of the unions in the workplace.

As part of this struggle over control of the production process, the company developed the so-called 'Project X' which was intended to prevent the unions from being able to stop production. The central tactic involved in the project was to introduce new technology into the workplace which could be operated by company-trained non-union labour. During a dispute in 1973, provoked by the dismissal of a print worker, the company implemented Project X. The company produced a special forty-page edition of the paper without the aid of striking typographers, who had until then been central to the production process, however, it was not published as the press operators seized control of the press rooms in a display of union solidarity. Nevertheless, the company had shown that it could prepare a newspaper without traditional craft labour, and in 1974 the typographers signed a contract which allowed the *Post* to introduce new technology as it saw fit, while providing current employees with lifetime job guarantees and a buyout provision. It soon became apparent, however, that management would not hire anyone to replace the workers who had been granted lifetime job guarantees. Indeed, as the new technology made craft workers superfluous from the perspective of management, the contracts of those workers were bought out by the management which was seeking means both to increase profits and to reduce union power.

Emboldened by the success of their dealings with the typographic workers, and by the subsequent introduction of cold-type technology into the composing room, management at the *Post* targeted the press operators whose contracts were to expire in 1975. Intending to regain control of the pressroom, the paper put a series of demands to the union, including reduced employee levels and increased management control over work scheduling. The response of the press operators was to go on strike, and in the course of the first night of the dispute a number of presses were damaged. Eventually, the company succeeded in producing the newspaper without the unionised workers, by using their trained non-union labour and by inducing journalists of the newspaper to cross the picket lines. With the use of the new cold-type technology, the newspaper company was able to produce the paper with a workforce of between 210 and 375, where it had previously employed 1220 craft workers. One step which made this possible was to contract much of the printing of the paper out to six other newspapers in the region. The company even 'hired helicopters which would pick up photographs of the pages of a twenty-four-page paper on the roof of the *Post* and fly them to these six points' (Kaiser 1985: 37). Management also successfully introduced cold-type technology to replace the existing hot metal production technology. As the dispute continued, the new production workers came to regard the jobs as their own, and made it clear to the management that they resented the possible return of the unionised workers. The management of the *Post* then presented the striking press operators with a list of demands which would give management the capacity to decide staffing levels and work schedules in the press room, thus removing the basis of union influence in the workplace. When the union rejected these demands, the company announced it would hire temporary and then permanent replacement workers for the press operators. Unity among the production worker unions also began to collapse as strike funds diminished and as inter-union rivalries arose. After new press operators were hired successfully, the Mailers' Union decided to return to work to be followed soon after by the other unions. The ultimate consequence of the dispute was that the press operators' union was decertified at the *Washington Post,* while reorganisation of the workplace occurred under the direction of management (Kaiser 1985; Smith 1980; Zimbalist 1979).

A number of important lessons for the newspaper industry in the US arose from this dispute, and were to be used again in later conflicts. Being able to draw on unskilled and alternative sources of labour to produce and distribute the newspaper using new production technology was vital for the bargaining position of the management. The production problems related to the withdrawal of labour by unionised workers could now be overcome by a management willing and able to employ a different workforce. This was to be a particularly important lesson for newspaper proprietors who controlled more than one newspaper. In such instances, the proprietors could bring in workers from other publications to produce the newspaper. The management practice of granting lifetime job guarantees in return for gaining control of the staffing of the workplace was to become another central feature of newspaper workplace bargaining in the 1970s and 1980s, and constituted a fundamental step in the movement of workplace influence towards management. Such developments also raise the question of the importance of a unionised workforce. While the newspaper industry in the US has been heavily unionised, it also has had many and various unions representing the workforce. Although these unions have been able to present a unified front in many disputes, the existence of ten or twelve unions in the workplace for much of the post-war period meant that the maintenance of solidarity could be difficult, especially as a dispute extended over time. In the case of the *Washington Post* this situation did arise as various unions came into conflict at different times. Previous animosities, personality clashes and links between management and union leaders and issues of race were to severely challenge the unified front that the unions needed to maintain if they were to succeed. Management was then able to use these inter-union conflicts to its own advantage by inducing some members to return to work. Without solidarity in the face of new technology, the position of the unionised workforce in the newspaper became increasingly precarious.

Since the events at the *Washington Post* heralded on a national scale the introduction of computer technology into the newspaper workplace, there have been many new developments in newspaper technology. A newspaper continually at the forefront of technological innovation has been the *Wall Street Journal*. In 1976, the newspaper began to use satellite transmission of its pages made up in New York; in the 1980s the paper used satellite transmission to expand worldwide; and by 1994 it was producing four various editions in eighteen US printing plants, with editions in Asia and Europe (Emery, Emery with Roberts 1996). At its production plants, management has attempted to introduce flexible workplace relations on the shopfloor with movement between craft areas expected. Available production technology provides management with the ability to remove traditional newspaper craft demarcations. In addition, at new plants the company attempts to employ non-union labour, while also employing a relatively high percentage of part-time and casual labour, especially in the cleaning and distribution areas (SOGAT 1985). Another newspaper leading the way in the use of technology is the *San Jose Mercury News*. It is known for its early innovative use of on-line services, whereby it offers its full editorial content to consumers on computer, having subscribed to America On-Line as early as 1993. Readers who wish to explore stories in more depth can gain access to the on-line service, where there are stored details of reports which the paper has not published in hard copy because of space and editorial reasons. This practice has since been followed by numerous other newspapers.

In the newspaper industry in the United States, the advent of cold type and computer technology has provided an impetus for major workplace reorganisation. Even at newspapers where unions were very influential, as at the *Washington Post*, publishers have succeeded in weakening the position of unions. Reporting on the situation of newspaper unions in the US after a study tour Brenda Dean, then general-secretary of the British non-craft print union, SOGAT, commented that the 'impact [of computer technology] turned out to be so vast, so devastating, that all of us returned [to Britain] somewhat shaken by what we had seen' (SOGAT 1985:1). Linked to declining union influence in many newspapers, publishers have been able to introduce technology which reduces labour requirements even further. Even though unions still have input into contract negotiations, and can withhold labour, as occurred in San Francisco in 1994 and in Detroit in 1995, the effectiveness of such actions is drastically reduced by publishers who are able to hire non-union labour, or bring in labour from other newspapers (*Guild Reporter* 1994d, 1994g). These developments have been further supported by the activities of state institutions such as the administration, the courts and the NLRB. Particularly through the Reagan era of the 1980s, these institutions sought to remove what were perceived to be the constraints of collective bargaining on productive and profitable workplaces. As such, union and state support for collective bargaining was attacked, and the right of management to manage free from outside interference was elevated as the non-union industrial relations system expanded in influence throughout industry in the United States.

Conclusion

The relational model of workplace organisation suggests that it is crucial to analyse the activities of actors including employers, management, union, workers and the state if we are to understand the relationship between technological innovation and workplace reorganisation in the newspaper industry. Experiences in the newspaper industry in Britain, Australia and the United States reveal that similar technological challenges have confronted the actors. In each country, computer-based technology has been introduced that has been at least partially responsible for a shift in the balance of workplace power from craft unions controlling work processes and labour supply to management, and to a lesser extent to journalists undertaking a broader range of workplace tasks. Similarly, the workplace actors have all acknowledged the need to adapt to new technology, recognising that it could not be resisted indefinitely. However, the institutional and social conditions in each country have resulted in important variations in how the processes of reorganisation have occurred. In the United States, a context supportive of individual bargaining at the workplace, relatively free from government or state intervention and increasingly free from union presence in the workplace, has allowed employers to act unilaterally in many instances, with unions being reduced to rearguard actions. In Britain, the election of the Conservative Party to office, and its engagement in a process of institutional and societal reorganisation, provided a context in which Eddie Shah could take the first steps to reducing the influence of craft unions in the workplace. In Australia, initial developments at newspapers such as Fairfax suggested that workplace reorganisation might follow a similar path to that taken in Britain and the United States. Even in the Fairfax case, however, the existence of the conciliation and arbitration system allowed unions at least to have a voice in disputes over technology and workplace reorganisation. This voice was to be strengthened in the 1980s through the Accord, and although major reorganisation of the newspaper workplace has also occurred in Australia, the intervention of the state has meant that at least in some instances this reorganisation has occurred in a context of negotiation mediated by the state. These similarities and differences in the relationships between technological innovation and workplace reorganisation will be examined further in part two of this study, which examines workplace reorganisation within News Corporation's newspaper holdings in Britain, Australia and the United States from the 1970s to the 1990s.

PART II

Technological Innovation and Workplace Reorganisation: News Corporation

CHAPTER 4

News Corporation Limited: A Global Media Company

News Corporation Limited has been referred to variously as 'the most global of all communications companies' (Koschnick 1989:98), and as 'the largest media group in the western world' (Advertiser Newspapers Limited 1986:20). Its Chief Executive, Rupert Murdoch, has been described as 'the most conspicuous of the high rollers in the media acquisition business [in the 1980s and 1990s]' (Emery, Emery with Roberts 1996:594), and as 'the man who would buy the world' (*Mail and Guardian* (South Africa) 1995:22). Analyses of its media holdings certainly support the notion that News Corporation is a truly global media company. In 1998, News Corporation, which has its registered office in Adelaide, South Australia, and its head office in Sydney, New South Wales, owned newspapers in Australia, New Zealand, Fiji, Papua New Guinea, the United Kingdom and the United States of America. Earlier in the 1990s, News Corporation had owned papers in countries as diverse as China and Hungary. The company also had global holdings in magazines and inserts, book publishing, television, filmed entertainment, satellite television, commercial printing, and various other operations including transport, computer and electronic technology, and music. In the financial year ending 30 June 1998, revenues of News Corporation totalled US$12.8 billion, an increase of 14 per cent on the previous year, while operating earnings grew to US$1.22 billion, an increase of 21 per cent (Murdoch 1998). Table 4.1 provides details of the revenues earned by News Corporation in its three major areas of operation, the United States, the United Kingdom, and Australasia.

In 1998, 74 per cent of the revenues of News Corporation came from the United States, 16 per cent from the United Kingdom and Europe, and 10 per cent from Australasia (News Corporation Limited 1998). In

Table 4.1   Revenues of News Corporation Limited for 1994–98 at 30 June

Revenues	1994 US$m	1995 US$m	1996 US$m	1997 US$m	1998 US$m
United States	5,544	6,283	6,535	7,837	9,489
UK	1,360	1,563	1,736	2,077	2,097
Australasia	1,081	1,147	1,281	1,302	1,255
Total Revenues	$ 7,985	$ 8,993	$ 9,552	$11,216	$12,841

Source: News Corporation Limited 1998

terms of operating income in 1998, 72 per cent came from the United States, 23 per cent from the United Kingdom and Europe, and 5 per cent from Australasia.

Throughout the 1990s, the company continued to expand its media operations by purchasing newspapers in countries including China and Hungary, while extending its cable and satellite television operations into countries such as India, Indonesia, Japan, Germany, Brazil and other South American countries. News Corporation has recently entered the arena of sport. It developed a new rugby league competition in Australia which involved buying the contracts of many of the best players in the world; established a European rugby league called Super League; purchased control of rugby league in New Zealand and other Pacific countries including Fiji and Manu Somoa; and purchased television rights for international rugby union in the southern hemisphere. Their satellite television channel, British Sky Broadcasting, signed a major television deal with the English Association Football Premier League, while the company was also involved in a bid to purchase Manchester United Football Club. In the United States, News Corporation owns the Los Angeles Dodgers baseball team, and is a minority owner of the New York Knicks basketball team.

It is perhaps not surprising that given all this activity, News Corporation has been the subject of much analysis. While not as large as other media companies such as Time Warner and Disney, the global reach and multi-media dimensions of News Corporation make it a crucial participant in international media. While it is evident that News Corporation has become a global media company, Rupert Murdoch has at times downplayed the notion that this has been the result of a deliberate global grand strategy on the part of the company. He has claimed, for example, that media do not operate at the global level but nationally and locally (1989b). At the same time, however, Murdoch does recognise that there is a global dimension to the company, and that global communications networks are a defining feature of contemporary media business practice. In responding to innovations in communications technologies, Murdoch puts the success of the company down to its capacity to take advantage of opportunities as they arise. Rather than an overall plan of action, News Corporation has been able to adapt to technological developments (1989b).

In the 1990s, this awareness of the importance of the global dimension has become more pronounced. In the company's 1998 *Annual Report*, Murdoch wrote that the company was intent on becoming a truly global international company, extending its reach beyond English-speaking countries. In developing this strategy, Murdoch went on to claim that News Corporation had an advantage over its competitors in that both the company and its management were international in their make-up (Murdoch 1998).

One indication of this globalisation was the claim made by the company in 1997 that soon more than 75 per cent of the world's population would have access to the programming platform of News Corporation (News Corporation Limited 1997). Central to the company's expansion has been its ability to seize on technological developments. Indeed, for Murdoch, innovations in information and communication technologies ultimately will mean much more than simply supplying new ways of providing and consuming entertainment and news. What is occurring through these technological innovations is a transformation of society in which wealth creation will be based not in industrial production, but in information processing (Murdoch 1989b). According to Murdoch, the dominant countries of the future will be those that can turn information into knowledge, by adapting to and controlling ideas and the opportunities offered by technologies in the information society (1989b).

From Murdoch's perspective, media corporations such as News Corporation clearly have a primary role in such possible societal transitions. While Murdoch is here talking about all forms of information and communication technologies, and is concerned with how they apply to all forms of media, newspapers remain particularly special to News Corporation. Newspapers have always been at the centre of the development of News Corporation, and the founding and subsequent globalisation of News Corporation is related to the success of its newspapers. Indeed, newspapers have been referred to by the company as the financial and cultural backbone on which the company was founded (News Corporation Limited 1998). Beginning in Australia with the *News* in the 1950s, an afternoon newspaper in Adelaide, Rupert Murdoch expanded his newspaper holdings into Perth, Sydney and Brisbane, and then nationally through the *Australian*, based initially in Canberra. News Limited became the dominant newspaper company in

Australia in 1986 when it purchased the newspapers of the rival Herald and Weekly Times group. The company first rose to international prominence before this through its foray into the British national newspaper industry with its purchase of the *News of the World*, and then famously with its purchase and overhaul of the *Sun* in the late 1960s. As it extended its holdings in Britain, the company also entered the United States buying newspapers in San Antonio in 1973, followed by further purchases, and subsequent sales, in New York, Boston and Chicago. It has since expanded its ownership of newspapers into other countries and other continents.

Newspapers also continue to be significant financially for the company. In the year ending 30 June 1998, newspapers provided 25 per cent of the company's operating income and 20 per cent of its revenues. Only television, with 35 per cent, provided more in operating income, while filmed entertainment (31 per cent) and television (26 per cent) provided more in revenues. Newspapers earned the company US$452 million in 1997–98, and throughout the 1990s the company has owned well in excess of 100 newspapers on an annual basis.

Coverage of the newspaper interests of News Corporation has focused on issues including journalistic standards, the links between Murdoch and political leaders, the company's competitive strategies, and the managerial approach of the company to a range of workplace issues, including technological innovation and workplace reorganisation. While much of the analysis of the journalistic quality of the newspapers owned by News Corporation has been critical of its standards – in particular in relation to tabloids – individual journalists at newspapers of the company continue to win awards (Auletta 1997; Coleridge 1993a, 1993b; Franklin 1997). There is a recognition that the support of Murdoch-owned newspapers has been crucial to the political success of leaders, ranging from the *Sun*'s support of Margaret Thatcher in Britain to the *New York Post*'s endorsement of Ed Koch, who was to become mayor of New York City. In other words, the newspapers are politically influential. In terms of competitive strategy, News Corporation has used price-cutting as a major weapon in trying to secure greater circulation involving newspapers such as the *Sun* and *The Times* in England, the *New York Post*, and the *Australian* (Glaberson 1994).

The management practice of News Corporation in regard to its newspapers has long been identified as being very much focused around Murdoch, who is a deeply involved and knowledgeable owner. While local executives throughout the world are able to make day-to-day decisions, Murdoch is seen as being very much in control of the organisation's operations. According to Richard Searby, one-time chair of

News Limited, management at News Corporation is very devolved and yet Murdoch still exercises great control over the decision-making of the company (Neil 1996). Andrew Neil, former editor of the *Sunday Times*, expanded on Searby's observation by suggesting that:

> [f]or much of the time, you don't hear from Rupert. Then, all of a sudden he descends like a thunderbolt to slash and burn all before him. 'Calculated terror' is how one of his most senior associates has described Rupert's management style – that and a simple but superb weekly financial-reporting system combined with an eerie grasp of numbers explain why he is able to keep tabs on almost all that is happening (Neil 1996:198).

According to Ken Auletta, a leading analyst of media owners, 'Murdoch's News Corporation is a large company run like a small one. A single individual makes big decisions, quickly, and for reasons he sometimes does not explain' (1997:275). In an interview in the early 1990s, Murdoch commented that the challenge he faced was to strike a balance between allowing editors to operate independently and taking responsibility himself (Coleridge 1993a). As part of this process, Murdoch claimed that on a day-to-day basis he allowed executives at the local level to deal with problems as they arose, although he would certainly intervene in a crisis situation. While Murdoch allows local managers room to operate, there is little doubt that he exercises overall control over the operations of the company.

The role of managerial practice in News Corporation has been very closely linked to the relationship between technological innovation and workplace reorganisation, and in this regard the organisation's news-papers have been at the forefront of defining the relationship. As will become apparent from the case studies to follow, the approach of the company to technological innovation and workplace reorganisation in the newspaper industry has been both revolutionary and controversial. The possible influence of technology on the future of newspapers was reflected in Murdoch's comments in a *Forbes* interview that the print media had no option but to adapt to the changes wrought by tech-nology. Without adjusting to the new technologies, newspapers would not be read, signalling their ultimate demise (Murdoch 1989b).

Newspapers, and the changing method of their production, have remained central to the operations of News Corporation, and have been crucial in its rise as a dominant global media group. Murdoch's belief in the significance of newspapers is apparent from his 'Chief Executive's Report' in News Corporation's 1997 *Annual Report* in which he wrote that print was still crucial to the company, even with the advent of the digital age. In particular, the managerial skills required in newspapers related to editing, creating content, and building a readership market would remain relevant to all forms of media into the twenty-first century (Murdoch 1997).

In the case studies that follow it will also become evident that the managerial skills required to run a newspaper involve the ability to organise a workplace, and are influenced by the history of specific workplaces and their location in certain institutional and societal contexts.

Case Studies

International comparisons would . . . be useful in highlighting the influence of political and institutional variables on the implementation and consequences of technical change [in the printing industry] (Frenkel 1990:61).

Although many studies already exist on the social impact of technological change, there seems to be a relative scarcity of studies on the reciprocal influence of labour relations and technological change (Ozaki 1992:1).

[T]he impact of technological change on labour can be examined most fruitfully within a framework which not only takes into account the specific characteristics of the new technological regime at the micro-level, but, at the same time, is capable of incorporating those pertinent political and economic relations in production and society as a whole (Leandros and Simmons 1992:41).

To examine relationships between the introduction of new technologies and workplace reorganisation I examined newspapers in three countries. The newspapers examined in depth were the Adelaide *Advertiser*, which is owned by News Limited in Australia, *The Times* of London and other British publications of News International, and the *New York Post*, owned by News America in the United States. I also undertook a more general analysis of the newspapers of News Corporation in these countries. In addition, I conducted field work at the *Financial Times* of London. While this is not a News Corporation newspaper, I was interested in analysing whether technological innovation and workplace reorganisation at News Corporation had any influence beyond its own newspaper interests. The *Financial Times* undertook major technological innovation and workplace reorganisation in the immediate aftermath of innovation and reorganisation at News International in London which suggested that the *Financial Times* would be an interesting comparative case study.

News Corporation Limited was selected as it is an organisation committed to substantial technological investment and workplace reorganisation in the newspaper industry. By undertaking a comparative study of newspapers within the same highly innovative organisation, it was anticipated that the primary industry-related factors influencing relationships between technological development and workplace reorganisation might be identified. Newspapers in three countries were chosen in an attempt to assess how differing national institutional and societal relations affected the relationship between technological innovation and workplace reorganisation in the newspaper industry.

Contemporary sociological theory on work and workplace relations indicates that the globalisation of technology use is having a substantial impact on workplace relations and practices. It is likely, therefore, that international convergence in technology use and in workplace reorganisation is occurring in the newspaper industry. By focusing on News Corporation Limited, an organisation involved in global expansion, it was considered that evidence might emerge as to global management strategies in regard to technology use and workplace reorganisation. Such analyses might indicate the extent of global patterns of resistance or accommodation to such technology use and workplace reorganisation on the part of the employees of the organisation, and the organised representatives of those employees.

From my base in Boston, I made field visits to the Adelaide *Advertiser*, and to News International and the *Financial Times* in London. In Australia, I interviewed members of the Printing and Kindred Industries Union (PKIU) who were active at the time new technology was introduced into the workplace in the 1990s and earlier at the Adelaide *Advertiser*. I also interviewed representatives of management, the editorial staff and the personnel department. In addition, I was taken on site visits of the premises of the Adelaide *Advertiser*, which involved observing the creative, production and distribution processes involved in producing and distributing a newspaper. During these observations, I interviewed journalists and print workers at their place of work.

In England I interviewed participants in the process of technological innovation and workplace reform. This involved interviewing management representatives of both News International and the *Financial Times*, in addition to journalists, print workers, and their union representatives. I conducted interviews, and observed work relations and practices at production and editorial sites of News International and the *Financial Times* at Wapping, Docklands and Southwark Bridge in London. I also visited the Trades Union Congress in London, and the headquarters of the Graphical, Paper and Media Union (GPMU) in Bedford. In addition, I was invited and attended a march and demonstration to commemorate the ninth year of newspaper production by News International at Wapping. At each site in Australia and Britain, management and workers' groups were extremely generous in providing documentation about the relationship between the introduction of new technology and how workplaces were reorganised.

My research on the newspaper industry in the United States was primarily documentary, consulting publications of trade unions in the US newspaper industry, including journals and newspapers, and management publications, including those of the American Newspaper Publishers Association (ANPA). I attended seminars on the US newspaper industry at The Joan Shorenstein Center at Harvard University, where I was able to speak to actors in the US newspaper industry, and I drew extensively on articles from the print media concerning technology and workplace relations in the US newspaper industry.

Conclusion

In the following chapters, case studies that examine relationships between technological innovation and workplace reorganisation in the newspaper industry are presented. Each case study seeks to highlight what is specific to the process of technological innovation and workplace reorganisation at each setting, while also identifying common developments across the cases. In the case studies, the proposition is examined that the relationship between the introduction of new technology and the reorganisation of workplace practices is dependent on the balance of power between trade unions and workers, management and employers, and the state, and that this balance of power must be situated within a specific set of institutional, political, social and economic relations.

CHAPTER 5

News International and Wapping

The History Of News International In Britain

By 1998, News International, the British-based company which formed part of News Corporation Limited, held extensive newspaper interests in Britain, as can be seen in Figure 5.1. Its newspaper holdings included the daily newspapers, the *Sun* which had a daily readership in 1998 of more than ten million, and more than 50 per cent of the daily tabloid market; and *The Times* which had an average circulation of 793,000 and a readership of more than two million. It also owned two Sunday newspapers, the *News of the World* with sales of 4.3 million and an average readership of 11.5 million, making it the leader in Sunday newspaper circulation, and the *Sunday Times* which had a circulation in excess of 1.3 million each week, and 48 per cent of the quality Sunday market (News Corporation Limited 1998).

In addition to these newspaper holdings, News International owned various magazines and inserts published in Britain including the influential *Times Supplements*; and book publishing interests through its ownership of HarperCollins Publishers. News International had also become a leading company in the area of satellite and pay service television, with 40 per cent ownership of British Sky Broadcasting and total ownership of a number of satellite channels. These holdings have made News International a crucial and immensely influential organisation in the media sector of Britain. The influence of the company, however, has been more widespread than merely securing a large share of the media market. In particular, the approach of the company to technological innovation and processes of workplace reorganisation has been significant in shaping the current state of industrial and workplace relations in Britain.

The history of News International's involvement in Britain began in 1969, when Rupert Murdoch arrived in England from Australia and

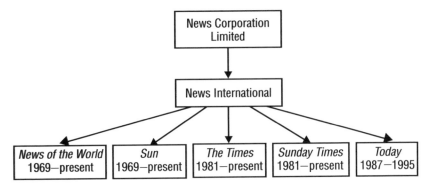

Figure 5.1 Newspaper Holdings of News International

bought the *News of the World*, which was the first newspaper to become available on the market after his arrival. This was followed soon afterwards by the purchase of the *Sun*. Development of the *Sun* from a daily newspaper with a circulation of less than one million to a newspaper which sold over three million copies on a daily basis, was a major success for News Corporation. In the following years, the company succeeded in reducing the size of the workforce at the *Sun* through agreements with the newspaper unions. Such reductions were, however, associated with work stoppages at the newspaper, as unions sought to either sustain or improve working conditions for the remaining workforce. The stoppages were also linked to the potential threat of technological innovations, such as direct inputting of newspaper text, which had the potential to remove the National Graphical Association (NGA) and its members from the production process (Jenkins 1979; Shawcross 1992a).

A further development which increased the influence of the company within the British media industry occurred in the early 1980s, when News Corporation acquired a significant shareholding in Reuters, the international news information service. In Murdoch's opinion, Reuters was a key element in the revolution in information (Shawcross 1992a). By acquiring an interest in Reuters, News International was able not only to gain access to key information resources but to benefit financially. This financial benefit arose with the increase in the value of Reuters' shares which accompanied its diversification into the global financial market, where it provided an electronic network to transmit stock and bond prices of currencies. The result was a surplus cash flow for News International, which made possible the purchase of new equipment and facilities, as well as other newspapers (Littleton 1992). In 1981, News Corporation bought *The Times* and the *Sunday Times* for

£12 million (Littleton 1992; Shawcross 1992a). Immediately, the new ownership sought to reduce the size of the workforce by entering into negotiations with the print unions which had favoured News International in the bidding for the newspapers. The unions had been supportive, as they considered Murdoch to be a tough but fair bargainer, and the eventual success of the *Sun* had meant jobs for union members. In 1987 the company purchased the newspaper *Today*. This purchase occurred, however, only after the traumatic move of the production and editorial site of News International from Fleet Street to Wapping. It is the dispute that accompanied this shift which is the focus of this chapter.

The Dispute At Wapping

The history of News International in the British newspaper industry has been one of expanding ownership and influence, and of challenges to the relationship between technology and workplace relations. For the company and its workforce the defining moment occurred in 1986 when production and distribution of News International newspapers began at Wapping in the London Docklands, away from the traditional base in Fleet Street. This shift in location resulted in perhaps the most significant development in the production processes and workplace relations in the history of national newspapers in Britain. Certainly, those involved in the developments of the 1980s were convinced of its importance. A senior member of the NUJ at the time talked, for example, of the move to Wapping as 'the point at which the industry was transformed', while Charles Wintour, an editor and proprietor on Fleet Street, wrote, '[I]n my view it was only through the outright defeat of the unions at Wapping that the great liberation of the British newspapers was finally secured' (1989:236).

The foundations for the dispute at Wapping were established in 1978 when News International purchased land at Wapping in the Docklands area of London. Through the late 1970s and early 1980s debts accrued by News Corporation from various ventures, particularly in the United States, meant that the company began to focus its activities on recouping money in other areas of its activities; one of these being the British newspaper sector. The company in England had succeeded both in reducing employment levels and in introducing technology into some of its newspapers. Attempts to introduce technology and workplace practices which would significantly restructure the company, however, had been rejected by the unions, which opposed redundancies they considered to be introduced primarily or solely in order to increase the profits of the company. The history of workplace relations in the British newspaper industry meant that such opposition from the unions was a significant factor to be heeded by management in its planning proposals.

From 1982, however, a new plan for company development began to emerge when a senior management member convinced Murdoch to build a new production plant outside of Fleet Street. The initial intent was that the new plant would be used only to print the *Sun* and the *News of the World*, the large circulations of which it was claimed required new production facilities unavailable in Fleet Street. Subsequent negotiations with the unions, however, did not achieve any agreement over the relocation issue. In particular, the unions were concerned with company plans to reduce the size of its workforce by up to two-thirds in the move to the new plant, and with company proposals for loosening job demarcations. The eventual result was that in 1985 News International developed the idea of moving to the new site without the involvement of the unions.

The plant itself had been completed in 1984 at a cost of £72 million, and the possibility of using an alternative workforce began to develop. Such a possibility became realistic after the Electrical, Electronic, Telecommunications and Plumbing Union (EETPU) signed a single-union deal with a Finnish paper manufacturing plant established in North Wales, and after it provided the workforce for the *Today* newspaper, established by Eddy Shah. The willingness of the EETPU to enter such agreements was a further stimulus for the company to introduce new technology into its operations, as it provided the opportunity of reorganising the workplace without the traditional print unions. Nevertheless, at this time, the company continued to enter negotiations with the print unions, although little progress was made. On 30 September 1985, members of News International, including Murdoch, met with the print unions to discuss the plant at Wapping. At this meeting, the company proposed plans for the publication of a new newspaper to be entitled the *London Post*, to be produced at Wapping, and it stated that discussion on the move of the already existing titles to Wapping would be resumed only if sufficient progress was made concerning the new paper.

Weeks later the company presented the print unions with a list of final demands, which it regarded as essential for the workforce of the *London Post*. The demands were that there was to be: a legally binding collective agreement; a no-strike, no lock-out agreement; no closed shop; and that management was to have the right to manage. Although such demands conflicted with all previous workplace practice in the industry, the print unions attempted to make some compromises. The print unions, however, demanded guarantees of jobs for life, and guarantees about wage fixation. These demands were rejected by the company, and in late December of 1985 it called a halt to negotiations with the print unions. The company continued with their plans to commence production at Wapping, claiming that the shift would occur whether or not agreement was reached with the print unions. As part of this plan, the company continued negotiations with the EETPU seeking means of establishing a single-union, no-strike deal. In response to these developments, the print unions held ballots in January 1986 to ascertain whether their members favoured taking industrial action. While waiting for the results of the ballot, and wanting to recommence dialogue with the company, the print unions made an offer to the company. The terms included:

1 Recognition of the right of the company to manage, and the right of trade unions to represent members.
2 Staffing to be determined by the company in consultation with trade unions.
3 Joint bargaining at plant level.
4 Binding arbitration on disputes, to be triggered unilaterally by one party.
5 Agreement that a mechanism was needed to avoid action which interrupted the continuity of production.
6 Flexible working, including use of new technology.
7 Single status for all employees with common conditions (Potter 1988).

As Potter suggested, these 'last ditch proposals' only a few months before 'would have seemed unbelievable' (1988:108–9). By agreeing to a single status for employees and to flexible working practices, for example, the print unions were indicating a willingness to give up job demarcations which had historically been central to their method of organising workplace relations. In addition, the proposals indicated that the print unions took seriously the threat posed by the company's plan to move to the new site without the print unions if agreement between the parties was not reached. Despite the intervention of the TUC in these negotiations, News International rejected the proposals outright, and on 19 January the company successfully produced and distributed, with EETPU and Transport and General Workers Union (TGWU) labour, 95 per cent of a twenty-four-page special advertising supplement to the *Sunday Times* at Wapping. On 24 January the national executives of the print unions, having received the results of the union strike ballots, called their members out on strike. The immediate response of the company, heeding legal advice, was to dismiss 5500 print and associated workers, and some journalists, for repudiatory breach of their contract of employment (Ewing and Napier 1986).

A dispute followed that lasted for over twelve months, capturing the attention of the national and international media, and involving peaceful demonstrations, incidents of physical violence, and the involvement of the coercive and parliamentary forces of the state, and which resulted ultimately in the complete move of News International from Fleet Street to Wapping.

In the course of the dispute, union members engaged in a number of tactics intended to stop the production and distribution of News International newspapers by alternative or 'scab' labour, that is, work undertaken in contravention of a union call for the withdrawal of labour from a workplace. The activities involved picketing and demonstrations outside the plant at Wapping; refusal to handle work proceeding to, or generating from, News International; and displays of national and international solidarity, including the boycotting of News International products. Print unions also called on the TUC to bring disciplinary action against the EETPU, whose members were being employed by the company to produce the newspapers.

The response of News International included a number of dimensions. While it portrayed the activities of the unions to be those of organisations opposed unreasonably to inevitable and necessary development, the company sought to downplay any effect of union activities on production and distribution. At the same time, the company turned to both the common law and legislation in an attempt to either prevent or limit the activities of the unions. The company successfully obtained injunctions restraining the unions from undertaking activity ruled to be obstructing the production and distribution of its newspapers, and also succeeded in obtaining the sequestration of the assets of the unions when the injunctions were ignored. In addition, the company received vital support from the police force. Indeed, Wintour has suggested that 'the ability of the police to keep the highway clear' was 'the most critical point' in the eventual success of News International (1989:222).

In addition to constraining the activities of the print unions, the company was able to maintain production and distribution of the newspaper. Previously when print unions withdrew their labour there was no alternative workforce immediately available to newspaper companies. The existence of a closed shop, specialised and craft skills, and union discipline and solidarity provided the workforce in the newspaper industry with a powerful bargaining position. In the Wapping dispute, however, this was not the case. The introduction of computerised technology meant that the craft skills of the print workers were no longer of central importance, and the company was able to draw on labour from the EETPU and train that labour virtually on the job. This strategy of the company depended fundamentally on the willingness of the EETPU to take the jobs of those workers engaged in dispute with the company. Further, despite strong opposition from the leadership of the union and from individual members, the majority of journalists, who were members of the National Union of Journalists (NUJ), crossed the picket lines and continued to work for News International. The company was also successful in using trucks and drivers supplied by the Australian-registered company Thomas National Transport (TNT) to distribute its newspapers by road. This meant that News International did not need to rely on the rail system for distribution purposes, where it would have encountered opposition from rail unions sympathetic to the print unions. In combination, these factors gave News International a very strong position from which to continue production and distribution of its products in the face of determined union opposition.

Even as the dispute was going on, print union leaders continued to meet with the company. Recognising that the company could successfully produce and distribute the newspaper, the tactics of the print unions at an official level became focused on trying to secure recognition within the Wapping plant. In all instances, however, they encountered a management determined that unions would not be recognised. The company did, however, make various settlement proposals to the print unions, which on each occasion were rejected by the union membership as being insufficient. Also, the company began to take a new approach by making offers to individual workers, independent of negotiations with the unions. Gradually the toll of daily activity in the dispute, and the threat of financial decimation through court actions and judgements, had an effect on the unions. In January 1987, the Society Of Graphical and Allied Trades (SOGAT) was advised that the latest court proceedings instituted by News International could result in the bankruptcy and financial demise of the union. In these circumstances, the National Executive of SOGAT voted to end the dispute unconditionally, and it was not long before the other unions involved in the dispute took similar decisions. Under these conditions, the company agreed to drop all legal claims in process, and offered a £58 million compensation package to dismissed employees. While unofficial protests continued for a short period, the increasing isolation of these attempts meant that they soon stopped, and the dispute at Wapping was over.

Wapping: A Dispute Over Technology or Workplace Power?

In considering the origins of the dispute at Wapping, the issues of technology, workplace organisation, and workplace relations must all be considered. Technological developments were central to the dispute. Wintour argued, for example, that 'key to the implementation of the [company's] plan was the computer typesetting system' (1989:219). Production of newspapers in Fleet Street was carried out using equipment that had long been made redundant, not only in other countries but in the regional areas of Britain. The technology introduced into Wapping was a US$10 million ATEX newspaper computer system which would be able to produce all four News International newspaper titles. Interestingly, the system was not considered to be at the forefront of technological development. The company wanted tried and tested technology with basic operating processes, at least in part because the workforce would not be fully trained at first (Wintour 1989). News International had made the decision to invest in this particular system after making a tour of newspapers in the United States in 1985 (Melvern 1986). The tour party consisted of News International representatives, along with a representative of the EETPU. In addition to observing the actual hardware, the members of the tour also noted the organisation of the workplaces involved. At *USA Today*, for example, it was noted that 'the editorial staff was responsible for typesetting and composition. Reporting staff also acted as copy-takers' (Melvern 1986:215). This blurring, or even abolition, of demarcations between jobs in the workplace was very different from the situation in Britain where union practice demanded the continuation of rigid demarcations. For News International, the technology being used in newspapers in the United States offered a possible way of reducing the influence of unions within the workplace. In the original plan for the new News International newspaper, however, there was to be no electronic page make-up; pages would still be pasted up. Over time, however, the company revealed that it intended to use full direct input technology in both editorial and advertising; at least in the proposed *London Post* newspaper (Melvern 1986). Also to be introduced into the new plant were Goss Mark One presses. As with the electronic system, these presses were not new. Indeed, according to Shawcross, Murdoch 'had bought them more than a decade earlier, when they happened to be for sale cheap. They . . . had been kept in their crates. Now they were literally dusted down and greased up and fitted with new electronics' (Shawcross 1992a:261). To complete the program of technological investments, the company purchased US$3 million worth of other equipment, including typesetting machines and graphics cameras, and purchased and developed computer equipment to handle data for the new distribution network of the newspaper (Littleton 1992; Shawcross 1992a).

For News International, this equipment represented a significant opportunity to reorganise workplace relations and to create more efficient and more profitable production. By allowing for a reduction in the size of the workforce, and by removing the need for many of the craft skills associated with print work, the new technology gave the company greater scope for flexibility in selecting members of its workforce. These developments would also make it more difficult for the workforce to take industrial action, as the threat of dismissal could be enforced by the company. The opportunity for a break from the history of relations between management and unions in the newspaper industry that this situation offered was addressed by Murdoch in 1985. He claimed that the constant opposition of unions on Fleet Street to the reorganisation of newspapers, in concert with their ability to disrupt production, meant that management was continually confronted with a choice of financial ruin or granting uneconomic wage and staffing demands. The result on Fleet Street was severe production department over-staffing and out-dated workplace practices (Potter 1988).

The connections between technology, workplace relations and profitability were further commented on, one year after the beginning of the dispute, in an editorial in *The Times* which claimed that the shift to Wapping had in itself freed the company from life-threatening financial losses, out-dated production methods, and 'anarchic trade union chapels' (*The Times* 1987: editorial page).

From such a perspective, the cause of the dispute was not technology or workplace reorganisation, but the obstructionist behaviour of the print unions in relation to new technology and workplace reorganisation. In March 1986, Murdoch claimed that the company had tried time and again to enter negotiations with the unions, but that the unions had refused, believing they could defeat him. The ultimate result of this belief, according to Murdoch (1986a), was the removal of the unions from the workplace. In confronting such union practices, the company considered that it had no option but to proceed as it did, which illustrates the importance of the history of workplace relations on the outcome of technologically-related workplace reorganisation. Previous protracted disputes were seen by the company as having been caused by union intransigence, and by 1986, News International had decided that a new way of bringing about reorganisation was necessary.

Union obstruction as the cause of the need for change also tended to be adopted in the mass print media coverage of the dispute. An article published in *Time* magazine, for example, portrayed the trade unions as being 'recalcitrant' dinosaurs who were holding up the advancement of technology and limiting the production capabilities in this sector of the manufacturing industry. As the *Time* magazine journalist stated:

for the past five decades, Fleet Street has . . . been synonymous with autocratic unions whose members operate the Linotype machines, produce the finished pages and run the presses. The unions have exercised a paralysing grip on Fleet Street, dictating who is hired, shutting down presses at will and, in effect, keeping the proprietors hostage (Kelly 1986:50–1).

While there was a grudging acknowledgment that management itself had contributed to some of the perceived problems confronting the newspaper industry, the majority of the criticism and vitriol was directed at the trade unions and their so-called 'industrial abuses'. Within that context, the introduction of new technology into the workplace was equivalent to the dawning of a new era for efficient workplace productivity and for increased profitability, which would benefit workers as well as management and proprietors. As Kelly indicated:

> For the journalists at Murdoch's Wapping plant, the sudden bypassing of Fleet Street's older and cruder ways is the dawn of the second industrial age. A few weeks ago they were crouched over typewriters, with oily presses thundering away in the pressrooms below. Now they sit in quiet offices in front of glowing screens on which they move paragraphs around electronically, rewrite clumsy sentences and call up their notes. As they go about their work, it is clear that there is no turning back (1986:56).

An even more celebratory tone was evident in *The Times*, which stated in an editorial entitled 'One Year at Wapping':

> A year ago, as well as the overmanning, the exploitative [workplace] practices and the interrupted production, there was a sense of alienation in our air. That has now gone. We trust it will never return. We cannot be complacent but we do feel more confident about ourselves. And, with the caution that must be ever by our side, we feel more confident about Britain (1987:editorial page).

For the management and ownership of News International, issues of technology and workplace reorganisation were presented as being about efficient and profitable production, and indeed about the continued viability of newspapers as a media form. The point had been reached where union opposition to development in the industry had to be defeated, and if this required confrontation with the unions then News International would follow that route. For the print unions and their members, however, issues related to the introduction of new technology and the transformation of the newspaper industry were linked fundamentally to the question of union power and influence in the workplace. Workplace control and the future of trade unionism became central issues for the unions, and for the rank and file members, as the dispute developed.

The official position of SOGAT in regard to technology at Wapping was expressed by the general secretary, Brenda Dean. She claimed that it is '[n]ever simply a case of do we accept new technology. We have to accept it. It is how it is negotiated that counts, that is the key' (Dean 1986a:29). In a speech delivered by Dean at a mass meeting of News International and *The Times* employees on 13 January 1986, she re-stated this view, arguing that 'SOGAT is prepared to accept new technology; SOGAT wants to go into Wapping and we do want to negotiate all the way there. We are not in a dispute of our making, it has been engineered deliberately by the company' (1986b:6).

A similar argument was made by Tony Dubbins, the General Secretary of the NGA, who claimed that:

> [t]he issue at Wapping is not about the use of new technology. The unions have positively responded to all of the technology demands which have been made by News International, and collectively agreed that the technology will be accepted . . . It is about whether or not trade unions in the national newspaper industry will exist and be able in the future to defend our members' employment, and advance their wages and conditions (1986a:3).

The position of the NGA was emphasised further by Dubbins when he argued in relation to the dispute at Wapping that:

> [i]t's not about new technology. The NGA and the other trade unions involved had earlier agreed to everything that Murdoch was seeking, including direct entry from editorial and advertising areas. In fact, some of the equipment at the so called 'hi-tech' plant at Wapping is far from new. For example, the printing presses are over fifteen years old. It's not about greedy Fleet Street workers wanting more money or better conditions either. It's about a blatant attempt by a multi national company to impose conditions on British workers which denies them the ability to defend their employment, wages, and working conditions (1986b:4).

The view that the dispute was about workplace influence and power, rather than about technology itself, was supported in a letter written by a rank and file SOGAT member after the dispute, in which he stated, '[I]n reality [the dispute] was a trade union struggle with political dimensions. This was clear throughout the dispute. Ballot after ballot turned down redundancy in favour of the right to work in Wapping' (Richardson 1988:10). Similarly, Neil Kinnock, then leader of the Labour Party, argued at a rally at Wembley (the Empire Stadium) in London that 'Stalag Wapping is about power and authority, not about technology' (1986:8). He added that if a system of management 'has to be packaged in concrete, wrapped in razor wire, and delivered by a huge police presence, then it really hasn't got much of a future as a system of industrial relations for Britain' (1986:8–9).

In seeking an explanation for the events that occurred at Wapping there are elements of the complete story in the positions taken by both the management and by the unions and their members. The technology introduced into Wapping, although not at the leading edge of technological capabilities, acted as an enabling device for the developments that occurred. It allowed the company to employ a different, and reduced, workforce which operated under a new mode of workplace organisation. At the same time, the resistance of the established workforce to the developments cannot be interpreted simply as being obstructionist. Union resistance was part of an ongoing struggle to ensure that members were not made redundant, but were given an opportunity to participate in the new workplace. In addition, the resistance was part of a wider struggle within British society over the future role of trade unions in industry.

Wapping: The Role of the State

Wapping and the Government

There is no doubt that the Conservative Party in office at the time of the Wapping dispute was hostile towards trade unions. Policies developed on the ideology of a free market and the supremacy of the individual, legislative enactments, and support for business in previous labour disputes made this apparent. Rupert Murdoch (1989a) was acutely aware of this particular situation, as he revealed in a speech in which he argued that in planning the move to Wapping the company had recognised the importance of the government's victory in the miners' strike, as well as the significance of labour law reforms, including protections against picketing. The political alignment between Murdoch and Prime Minister Thatcher was reflected on by Charles Douglas-Home, who was to become editor of *The Times* under the ownership of News International. Douglas-Home was reported as saying that 'Rupert and Mrs Thatcher consult regularly on every important matter of policy . . . especially as they relate to his economic and political interests'. He went on to say that 'around here he is often jokingly referred to as "Mr Prime Minister". Except that it's no longer all that much of a joke. In many respects, he is the phantom Prime Minister of this country' (quoted in Kiernan 1986:309, 317; Littleton 1992). During the course of the dispute at Wapping, Thatcher made the direction of her sympathies plainly evident. In response to a letter from Robert Litherland, the Labour member of parliament for Manchester Central, concerning the dispute at Wapping, the Prime Minister replied that she was 'dismayed at the television pictures of violent picketing outside the [Wapping] plant'. She continued:

> I am afraid that the current problems of the newspaper industry are in large measure the result of resistance by the print unions to technological change and their desire to preserve very substantial overmanning. Our industrial relations legislation is perfectly fair and has restored the balance between management and unions which had been tilted so heavily in favour of the unions in the 1970s. It is of course open to anyone, whether management or unions, to have recourse to the law, and it is incumbent on all concerned to obey the rulings of the courts (Thatcher 1986:4).

Similar sentiments were expressed by other Conservative Party Members of Parliament during the course of the dispute. Norman Tebbit, Conservative Party Chair, wrote, for example, that:

> Like ostriches the printers of Fleet Street buried their heads in the sand. They hoped the new technology would go away and, more reactionary than any country squire, they rejoiced in having resisted change . . . Fleet Street has for years been littered with the financial corpses of the press barons who could not stand against the losses from producing newspapers the printers' way . . . but they must have known the type had been set a long time ago for this story. I am glad to say that Fleet Street in general now represents no more than an absurd anachronism in British industry (quoted in Littleton 1992:51).

In another instance of support, a senior government official was reported as declaring that 'Fleet Street is one of the great bastions of Luddism. The print unions, which have rejected every attempt to adapt to the future, are now washed up on a very lonely shore' (unnamed source, quoted in Kelly 1986:55–6). Further support for the company was expressed at the end of the dispute, when Mr Kenneth Clarke, the Minister for Employment, was reported as saying that the NGA's unconditional withdrawal from the dispute had shown that 'even the most hardline and militant unions had to comply with the law'. He went on to state that 'most people will be glad that the law has been upheld, and that even the NGA have at last accepted that these laws are there to protect public order and make sure disputes are conducted in a more sensible way' (quoted in Jones 1987:1).

Public pronouncements such as these made it evident that the unions involved in the dispute at Wapping could not expect support or sympathy from the government. Within this context, which was in marked contrast to the government–union relationship under the previous Labour Party government, the role and use of the institutions of the common law and legislation were to be critical to the outcome of the dispute.

Wapping and the Law

The law was used on a number of occasions in the Wapping dispute, often with devastating results for employees. In particular, the law was used by News International as one means of pursuing its objectives of introducing technology and restructuring the workforce. The company received legal advice that, at common law, any worker taking strike action could be dismissed immediately, and that such a worker would have no claim for either unfair dismissal nor for redundancy, given that the very action of striking constituted a repudiatory breach of the contract of employment. Crucially, from the perspective of management, the dismissals had to occur after a call for industrial action. If they occurred prior to such a call, then a claim for redundancy payment could well succeed (Ewing and Napier 1986; Potter 1988; Richards 1986). In addition, under Section 62 of the *Employment Protection (Consolidation) Act 1978*, various protections for striking workers had been removed in relation to the situation in which the strike precedes the giving of redundancy. While this legislation and the common law were in place prior to the Conservative Party gaining office, legislation enacted by the Conservative Party government made the 1978 Act more readily accessible. In particular, the *Employment Act 1982* enabled an employer to use Section 62 of the 1978 Act on a workplace-by-workplace basis, removing the need to act consistently across an entire multi-site organisation (Potter 1988).

The impact of the Conservative Party government legislation also narrowed many rights related to picketing, including the removal of immunity from prosecution under the common law for acts done in the course of picketing. Picketing is an important means by which unions can make public their grievances with a company, and develop a sense of solidarity among members. From very early on, pickets at and around the premises of News International, along with public demonstrations, were central to the tactics of the unions. Demonstrations were organised every Saturday evening, and also at peak times for the distribution of the newspapers from the plant. In addition, daily pickets were placed at the premises of the company, both at Wapping and Fleet Street, as well as on the road leading to the Wapping plant. At first, the company attempted to downplay the effect of such activities, claiming that although the pickets were vocal, they had not been effective in stopping the movement of delivery trucks or cars (Murdoch 1986a). The company did eventually take such activities seriously, bringing an injunction application against the unions and union officials. The action of picketing, and acts undertaken in the course of picketing, were found by the court to have infringed a number of torts related to acts done in contemplation of or furtherance of a trade dispute. Prior to 1980, these actions would have been immune from the law of tort precisely because they were undertaken in relation to such a dispute. The combination of Section 15 of the *Trade Union and Labour Relations Act 1974* and Sections 15 and 16 of the *Employment Act 1980*, however, removed this immunity.

News International was also able to use the Conservative Party legislation to seek injunctions to prevent unionised workers at non-News International workplaces from taking sympathy industrial action in support of the striking workers, and to prevent such workers from refusing to handle News International newspapers. While the use of law in industrial disputes was not new, in the 1980s 'the law came to play a much greater part in employers' strategies' (Elgar and Simpson 1993: 94). A further indication of the role of the law in the Wapping dispute was that the employer sought contempt proceedings, and attempted to sequestrate the funds of the unions when the initial injunctions were not obeyed. Indeed, some two weeks after the court initially issued an injunction against SOGAT, evidence was presented to the court that the injunction was not being followed. The response of the court was to order the sequestration of the assets of the union and to impose a fine of £25,000, accompanied by a warning that more severe fines would soon follow if the injunction was not obeyed. The NGA was only able to avoid a similar fate by apologising for its breaches of the injunction, having observed the fate of SOGAT. Contempt proceedings again threatened the unions in early 1987 when court actions for breaches of picketing injunctions threatened the unions with financial ruin. In terms of the dispute, the significance of these threatened injunctions was identified by Cockburn, who observed, 'When SOGAT and then NGA capitulated to Murdoch in February 1987 it was largely because the unions were in peril of financial ruin from court actions in pursuit of a dispute they no longer saw any chance of winning' (1991:241). Despite avoiding financial ruin, the effect of court actions on the unions was significant. For example, while SOGAT eventually had the sequestration order removed, by the autumn of 1986 an estimated 50 per cent of SOGAT's assets of £17 million had been spent on the dispute or lost in the sequestration of its assets, while it was estimated that possible damages awards could amount to £2.5 million (Potter 1988).

These events illustrate the importance of the interaction between workplace practices and the institutional regulation of the workplace. The company's determination to proceed with workplace reorganisation was facilitated by the existence of legal regulations supportive of its goals and restrictive of the actions open to the union. While the punitive dimensions of the law were in full evidence in the dispute, the unfolding of the dispute also illustrated the limited influence of government mediation bodies such as ACAS (Advisory, Conciliation, and Arbitration Service). This legislatively-created body offered to mediate and arbitrate during the dispute, only to have such offers rejected by the company which argued that any attempts at negotiation were bound to be frustrated by the unions (Littleton 1992). In these circumstances, the use of law that carried with it possible punitive measures was more appealing to the company.

The dispute at Wapping revealed many facets of the interventionist potential of the law in workplace relations in Britain. Enforcement of the law was to be a further vital dimension in the role of the state in the dispute, and a particularly prominent role was taken in this regard by the police force.

Wapping and the Police

An important element of the dispute at Wapping was the involvement of the coercive forces of the state – in particular, the police. While police have commonly been used to preserve the public law and order in industrial disputes, the levels of police involvement at Wapping were extraordinary. One estimate of the total cost of police activity in the dispute was £14 million. An average of 300 police were on duty at the plant every day of the conflict which lasted for more than twelve months, and on twelve occasions the number exceeded 1000. From management's perspective, there was good reason for the high level of police involvement in the dispute which, as has been noted, frequently involved mass picketing and very large numbers participating in demonstrations outside the News International plant. According to Rupert Murdoch, in a speech made in New York in 1989, a key element in the Wapping dispute was the role of the police in ensuring that company trucks could move in and out of the plant, in the face of pickets. For Murdoch, in fact, the ability to deliver newspapers from Wapping by truck was ultimately a decisive factor in the company's eventual victory (1989a).

While the police played a critical role in allowing for the successful distribution of newspapers, and hence in the ultimate success of News International, questions of violence on the part of the police force emerged with alarming frequency. A member of SOGAT who participated in marches wrote:

> I was proud to join my comrades in the peaceful demonstration at Wapping on Saturday 24th January 1987, but was sickened by the forces of democracy

in their manner of crowd control. It was even more galling to think I was helping to pay the wages for these people to bash my colleagues over the head – although it must be said, like the demonstrators, some were more enthusiastic than others. I was told by one that if I did not get on the pavement I would get his stick over my head. My opinion of the events, and I am 57 years of age, with no police record but a long history of peaceful demonstrating, was that the cavalry charge by the mounted police . . . followed by storm troops with drawn batons, was altogether a frightening experience (Donnelly 1987:10).

A report by the Police Monitoring and Research Group (PMRG) provided extensive details of police activities at the News International plant. Incidents on a number of different dates were documented, reporting violence perpetrated against picketers and demonstrators. Legal observers reported that on Saturday 10 May 1986, for example:

At about 12.30am a group of approximately 12 mounted officers equipped with riot helmets approached along The Highway from the east. At this point, 20 yards from the crowd, the pace of the officers increased to a trot and they rode towards and into the body of demonstrators who were trapped between the horses and the police line.
I saw people running in panic including a number of middle aged men and women who had difficulty getting out of the way. Several people fell to the ground in the stampede. Having ridden through the crowd the horses were turned and formed a picket in front of the original picket line. As they rode the final distance through the crowd I saw one or two missiles being thrown, most appeared to be plastic cups of coffee which people had been drinking.
There was considerable confusion as a result of this charge which appeared to have no strategic purpose whatsoever in my view. The arrival of the officers was greeted with a perceptible heightening of tension and a great deal of hostility (PMRG 1987:14–15).

Members of the Graphical, Paper and Media Union (GPMU) vividly recalled the role the police had played in the conflict. One member spoke of the 'battle of Wapping', recounting that 'we were walking as a group and there were police ahead of us, and behind us. We turned, did not run. If we had run, the police would have charged. We linked arms and walked; half our crowd disappeared. The back of my neck was covered in horse saliva, and when I turned around, I just saw a giant pair of nostrils'. She recalled further that 'riot police came in with batons and horses, and were whacking people left and right. We had eight speakers, and they only spoke for a total of ten minutes. It was horrible. People were crying – not just the women, also the men'. A former print worker also had vivid recollections of the police tactics used at Wapping:

One of the most noticeable things about the Wapping dispute – about the policing of it – was that the police actually selected the older people, clearly less militant; the ones that weren't shouting out. They singled them out and gave them beatings first. I'm quite sure that their strategy there was that the numbers will quickly reduce if we go for the weak ones . . . which happened. People who went down there really believing they were fighting a good fight suddenly found they were in the way of a police horse coming charging down on them with a truncheon, and they wouldn't go again. It was too dangerous . . . I know people that I worked with who went down there in the early weeks, and they were frightened off it.

By 23 January 1987, 1,370 people had been arrested, and 1,058 had been convicted, with 128 cases pending. Five hundred and seven people were charged with threatening and insulting behaviour, 298 with obstruction of the highway and 177 with obstruction of police. On being taken to the police station, individuals were searched, photographed, and asked details relating to their trade union membership (PMRG 1987). The police recorded that 574 of their members had been injured in clashes with demonstrators, while unofficially more protesters were injured, and more than 1000 attacks on News International- and TNT-affiliated vehicles and drivers were recorded (Littleton 1992). In the years following the confrontation, prosecutions were brought against a number of the police force involved in clashes with the public.

The dispute at Wapping revealed the influence that the state could have on the processes and outcomes of workplace reorganisation. Government support for the company and its hostility towards the unions, when combined with interventions from the courts and the police, were to be critical to the outcome of the dispute. However, the conflict at Wapping revolved ultimately around relations between the company and its workforce, and it is to the role of the unions in the dispute that attention is now directed.

Divisions Within the Trade Union Movement

Fundamental to the events that occurred at Wapping in the mid-1980s were the divisions that arose between and within trade unions and the trade union movement. The most notorious of these was the divide that arose between the EETPU and the print unions. The EETPU played a vital role in that it was willing to negotiate with News International about moving from Fleet Street to Wapping. In seeking to increase its membership base and provide jobs for existing members, the EETPU had entered into single-union deals with a number of employers in the 1980s, including Toshiba in 1981 and the Messenger Press in 1983 (Ewing and

Napier 1986). The significance of pursuing a similar policy at Wapping soon became evident and, according to Gennard, 'without the cooperation of the EETPU it would have been difficult, if not impossible, for News International to recruit an alternative work force to print its papers' (Gennard 1990:509). EETPU members provided the bulk of the production workforce at the new site.

In explaining the actions of the EETPU, the general secretary of the union, Eric Hammond, downplayed the idea that there was a deliberate strategy being followed, claiming, 'We've not got some grand new strategy. We do what seems sensible on a day-to-day basis. Out of that, a new type of trade unionism has developed ... It's just the way it's come out' (Hammond 1988:12). In relation to the newspaper industry, Hammond's perspective was that EETPU members should be able to develop their skills to take advantage of technological developments, especially in electronics (Melvern 1986). The result in this particular instance was that while the EETPU had no official agreement with News International, it played an active role in supplying workers, who were then employed on individual contracts.

For the NGA, the activities of the EETPU in regard to Wapping amounted to treachery. According to an NGA briefing paper:

> [T]he EETPU were quite willing to jettison any trade union principles they might have had in order to cynically collaborate with the employers in getting rid of the 6500 strong work force at News International in return for 'nod and wink' assurances of a 'sweetheart' arrangement at 'Fortress Wapping' (NGA 1986a:15.3).

The NGA briefing paper claimed further that '[w]ithout the complicity and collaboration of the EETPU, Murdoch's "grand design" would have failed' (NGA 1986a:15.4). According to the NGA, the treachery of the EETPU was heightened by the secrecy of their role in the move to Wapping, until finally in discussions the EETPU 'made it clear that they were not prepared to be part of a "joint approach" with the other unions', and that, furthermore, 'they could live with Murdoch's demands' (NGA 1986a: 16.15, 16.16). Other print unions put pressure on the TUC to take disciplinary action against the EETPU.

The response of the TUC was to issue a six-point directive to the EETPU, with the threat that the EETPU would be suspended until the Annual Congress of the TUC if it did not comply with the directives. The main thrust of the TUC directive was to order the EETPU not to enter into unilateral negotiations with News International, and not to recruit any further workers at News International sites without agreement from the other unions (NGA 1986a). These directives were not strong enough for those involved in the dispute, who argued that the

TUC should have demanded that the EETPU instruct its members not to cross the picket lines, and not to undertake work done by members of the print unions (NGA 1986a; SOGAT 1986). Feelings of hostility were heightened further when evidence arose of contacts continuing between the EETPU and News International despite TUC instructions.

Hostility towards the EETPU was also expressed by rank and file union members, as illustrated in an article written by the 'Sacked *Sunday Times* NUJ reporters' in 1986. They wrote of the situation at Wapping in the following terms:

> The dignity, credibility, and integrity of British trade unionism today hangs by a thread from the razor wire surrounding a fortified print works in the east end of London. The most cynical and devastating betrayal of the principles on which the trade union movement flourished is now entering its eight month behind the guarded gates of Wapping. At this moment members of the Electrical, Electronic, Telecommunications and Plumbing Union . . . are doing the jobs of 5,500 strikers ruthlessly sacked by Rupert Murdoch.
>
> . . . At the end of their shift, the Wapping electricians, presented to the world as the vanguard of the electronic revolution, will, like a million printers before them, head for the wash room coughing, spitting and nose-blowing 'ink fly' and paper dust out of their heads.
>
> Outside the gates of Wapping a betrayed generation of stereotypers, process artists, copy-readers, proof-readers, compositors, darkroom technicians, librarians, machine minders, cleaners, printer–planners, clerks, typists and maintenance engineers and their families watch in helpless fury as the great jobs swindle continues on the nod of the most hostile and paramilitary police force in Britain (Sacked *Sunday Times* NUJ Reporters 1986:1).

Concern with loss of jobs was one reason for the hostility shown towards the EETPU. Another was that the EETPU had broken union discipline and solidarity. The possibility of success for the unions in the dispute with News International depended crucially on the withdrawal of labour from the workplace, but actions of the EETPU meant that News International had access to a production workforce.

Divisions within the union movement during the course of the Wapping dispute developed not only between the print unions and the EETPU. Rivalries among the print unions themselves arose over issues including future staffing and organisational arrangements, and over control of the workforce. As Cockburn (1991) suggested, these rivalries could be traced to long-standing sectionalism based on the continuing demarcation in membership according to whether labour was classified as craft or as unskilled. An example of the tensions that were generated from this divide was provided by Brenda Dean (1985), who argued that although new technology was leading to changes in the work done by SOGAT members in the newspaper industry, this did not mean that other unions could start to claim that work, those jobs or those members.

More significant in relation to the dispute at Wapping, however, was the division that developed between the print unions and the NUJ on the one hand, and the rank and file members of the NUJ on the other. The official policy of the NUJ was to support the print unions in the dispute with News International (NUJ 1987). Although this call for solidarity was supported by a number of journalist members of the NUJ, many voted to continue working for News International at the new site. One explanation for this divide was provided by a journalist, Peter Wilby, who at first refused to move to Wapping, but then did so. In a 1986 interview, he stated 'Like most journalists on the Street, I dislike the print unions' (1986:31). While he professed a grudging admiration for the ability of the print trade unions to preserve the jobs of their members at high wages, he claimed that journalists resented the manner in which print workers had previously crossed the picket lines of journalists. They also resented the influence that the industrial actions of the print unions had on whether or not the work of journalists would appear. The options put to individual journalists by News International were also stark. If they moved to Wapping, they would receive an annual salary rise of £2,000 and free health care coverage, and 'would be expected to put copy directly into computer terminals, thus usurping NGA functions' (Wilby 1986:31). If they refused, they would be sacked. While some journalists were dismissed for refusing to go to Wapping, eventually the great majority of the 700 journalists on all four of the newspaper titles of News International did go to Wapping (Littleton 1992).

For many of those involved, whether or not the journalists would go to work at the new plant was the crucial issue of the dispute. According to one senior NUJ official, the whole move was a 'terrible gamble' on the part of the company, because it did not have the support of the journalists before the shift. While the company could recruit a new production staff to operate the presses, it would be more difficult to recruit a new journalistic staff with which to maintain or increase the circulation of the newspapers. Trying to explain the eventual decision of the majority of journalists to cross the picket lines, this union official recollected the events surrounding the decision to go to Wapping:

We had a meeting that went on most of the weekend at the end of which the *Times* journalists just decided that they wouldn't go [to Wapping] on the Monday, and that they would reconsider the position, and then a load of them just went. The thing about *The Times* and *The Sunday Times* chapel is that they were nothing so much as like the Oxford Union, where a lot of their members came from. It was all a great debate but, because it was only a debate, nobody felt bound by the decision.

Similarly, an NUJ father of the chapel observed that 'journalists as a bunch tend to be pretty individualistic. They were never subject to the absolute disciplines that print unions could exert over their members'. Given the history of relations between print workers and journalists, it proved impossible for the leadership of the NUJ to instil in its members sufficient solidarity for them to take industrial action which might result in job loss. As one journalist said, 'for most of us, although there wasn't much sympathy for Murdoch, there was a feeling that this hour [of technological transformation] was bound to come eventually, and that the journalists owed the print workers nothing, so they were damned if they were going to put their jobs at risk as well'. In these circumstances, the risk taken by News International to commence the move to Wapping without an agreement with the NUJ was ultimately justified.

Further divisions arose during the course of the dispute, including confrontation between the print workers and the union leadership, which was illustrated at the end of the dispute (Dubbins 1987). While the union leadership saw no other way of proceeding than by settling with the company to avoid bankruptcy, rank and file members argued that they had been betrayed by their leaders. The editorial in the final edition of the *Wapping Post*, dated 31 March 1987, for example, reacted angrily to the union leadership 'retreat'. Under the headline 'We will never forget', it was stated:

> The sacked News International printworkers, with the full support of their chapels and branches and despite the shock of the retreat by their National Union leaderships, have correctly resolved to continue the struggle against Murdoch, Hammond, the Tory anti-union laws, the multi-nationals and the oppressive state machine (*Wapping Post* 1987:6).

The writers of the editorial were outraged at what they considered to be a sell-out by their leaders to the company, and to the courts. These arguments were supported by Arthur Scargill, head of the National Union of Miners, who criticised the Trades Union Congress leadership, and lamented:

> Each time a new battle is launched against workers and their unions, we have rhetorical rallying cries from the official leaders of the trade union movement. How much longer though, can workers themselves give any credence to those cries unless they see leaders prepared and ready to mobilise effective action? If that means defying the Tory Government anti-trade union laws, then so be it. Those laws should be defied and we should accept the consequences (1987:6).

Divisions within the print unions were also evident as print workers who were not involved directly in the production of Fleet Street newspapers were sometimes unenthusiastic in their support for the dispute. This was shown most clearly when union ballots, held to raise a financial levy from all members to support those workers involved in the dispute, were defeated.

Divisions between and within unions meant that those on the labour side could not focus solely on their argument with the company. Particularly in the case of the EETPU, there was a high level of turmoil within the labour movement which was to be a severe constraint on the possible actions open to the union movement. The divisions also assisted News International. While the company was in conflict with print workers and their unions, a key difference from previous confrontations was that disunity in the union movement provided one means by which the company was able to recruit an alternative workforce, and to retain the majority of its journalistic staff. As a consequence, News International was in a very strong position from which to continue both newspaper production and distribution. It was also in a position to create a workplace at Wapping which was to be considered by other newspaper companies as a possible model for their own future workplace organisation.

Wapping in the 1990s: A Non-Union Workplace

The future of workplace relations, and of labour practices at Wapping, was predicted in a *Times* editorial of 19 January 1986. The editorial argued that Wapping would be a model for newspaper production of the future, with its up-to-date technology, new workplace practices and progressive workplace organisation. The editorial argued that despite numerous recommendations and attempts to reorganise newspapers in the past, a combination of union power and management weakness had prevented necessary change from occurring. It went on to say that reorganisation was now finally inevitable, but that the way that such reorganisation would come about was very much up to the unions. They could either cooperate and adapt to the new reality, or resist and 'risk losing everything' (Melvern 1986:246–7).

In terms of the organisation of work at Wapping, managerial practice since the dispute has involved various forms of human resources management. The basic elements include personal contracts for employees, no union recognition or representation, and advancement in the workplace on the basis of individual 'merit'. Rather than organising work by hierarchy and demarcation, there has been an attempt to introduce team work. These teams are self-monitoring, independent of external control, and carry the responsibility themselves for maintaining high standards (Littleton 1992). As part of this process, the company embarked on extensive retraining programs, both for shop-floor employees and for managerial staff.

Despite the advances that the company claims to have made, questions still remain over the modes of negotiation and workplace representation that exist within the plant. The developments at Wapping resulted in the derecognition of the unions and their subsequent absence from the bargaining process. While individual employees are not prevented from being members of a union, the company does not recognise unions and does not bargain or enter into negotiations of any sort with them. A human resources manager was reported as having explained, for example, that while any member of the workforce could be a union member, the company did not recognise unions and did not negotiate with unions. In short, members of the workforce would not get any benefit from belonging to a union (Littleton 1992).

An early indication of the implications of the new workplace arrangements was provided in November 1987 when SOGAT revealed that there were serious problems facing the workforce in Wapping. Reporting that workers had approached the TUC for guidance on union recognition, SOGAT claimed that internal documents revealed '[l]ow morale among the Wapping workforce, disillusionment with the electricians' union the EETPU and resentment against the style of management' (SOGAT 1987:8). Minutes of a meeting at Wapping in August reported that:

> the staff thought that since the end of the strike communications had deteriorated and the men's morale had dropped. They felt that the management now had an advantage over the workforce and nothing had been done to alleviate this feeling. Since the pickets had disappeared the management had come down more heavily on the staff and it made them feel insecure (SOGAT 1987:8).

In addition, a new style of bargaining over wages was revealed when a staff claim for a wage increase was ignored. The company imposed a 6 per cent wage increase, with the managing director claiming that this figure was non-negotiable (SOGAT 1987:8). In the new workplace, decisions concerning wages and work conditions were to be made by management.

There is, however, a human resources committee, comprised of equal numbers of management and staff, to deal with complaints and grievances. While this committee provides employees with a mechanism to express concerns, it is far removed from the influence exerted by the print unions in Fleet Street. Commenting on the role of management in a workplace described as being a 'constraint free environment'

(Oram 1987:89), a human resources manager explained that with the removal of restrictive workplace practices associated with the presence of the unions, management now had much greater flexibility to run the business more efficiently. In particular, the manager claimed that this was the case because there was no longer any need to enter into negotiations with unions and the chapels over production issues, which in the past had slowed down the ability of management to respond to business opportunities (Littleton 1992).

While the minimal level of union membership at Wapping is in large part due to the explicit derecognition practices of the company, management claims that derecognition has not had a negative influence on the workforce, and has actually resulted in an improvement in productivity and in workplace relations. According to members of the GPMU, however, the process of derecognition has involved the exertion of coercive forms of power, and that a critical factor influencing the low level of union membership is fear on the shopfloor induced by managerial practice. They have found that workers are afraid of facing disciplinary action or dismissal for union-related activities. According to one senior GPMU organiser and official, 'What happens is that each time [there is an attempt to gain union recognition] the people at the forefront are singled out and sacked'. Another official argued, 'Wapping is non-union for us. We have a few members there but they have to be there secretly, working in horrendous conditions. It will be at least four or five years before we can become more visible, and overcome the fear'. For the union movement, the current situation at Wapping was clearly summarised by an NUJ father of the chapel who suggested:

> [I]t's going to be a long while before you've got anything like full NUJ membership at Wapping. You've got to be able to convince people that there's a point in being a member. The union isn't recognised, management is required to be pretty hostile, and I can't see in Murdoch's lifetime that that is going to change very much.

While managerial practice at News International is clearly vital in explaining the levels of union membership at Wapping, these low levels have also been influenced by early union reluctance to organise in a workplace populated, at least initially, by workers who broke union solidarity by crossing picket lines. As a senior member of the NUJ observed:

> We went through a debate, an agonised debate, that lasted about three years, over what we should do with the members who had defied the instruction [not] to go to Wapping. Should we expel them? should we fine them? should we congratulate them? or ignore it? In the finish – quite rightly in my view –

we did next to nothing. But that was a huge debate that dominated a couple of conferences.

There was also hostility felt towards the EETPU which continued to be a presence within the workplace, despite TUC instructions that the union was not to continue contacts with people at the new plant.

While the company has resisted attempts to re-establish trade unionism in the plant, members of the print union are beginning to agitate for a union presence there, even though there is evidence of continued hostility towards those who went there in defiance of union orders. In 1990, delegates of SOGAT voted to start recruiting membership again at Wapping, reversing a decision taken two years earlier (SOGAT 1990). Billy Osborne, father of the chapel of SOGAT at *The Times* when the Wapping dispute occurred, wrote that 'We've got to be realists'. He went on to explain, 'Yes, we decry, we deplore what has happened to our members, but we must have some sort of recognition in what is going to be the biggest plant in this country and possibly in Europe' (Osborne 1990:25). Despite the recognition by unions of the desirability of organising at such an important company, in the face of continued company hostility towards any form of union presence at Wapping it has proven to be extremely difficult to achieve success in recruiting new members.

The Wapping Effect: The National Newspaper Industry After Wapping

Events at Wapping were to have immediate repercussions throughout the British national press. Within sixteen months of the move of News International from Fleet Street, all national newspapers had moved to new sites or announced plans to move, in most cases to the London Docklands. At newspaper companies such as the Telegraph Group, Associated Newspapers, Mirror Group Newspapers, and Express Newspapers, technological innovation and workplace reorganisation were integral elements in the relocation process. These developments were achieved by negotiation between management and the unions, so that unions maintained a presence at the new worksites. Such negotiations were carried out in a very different context from negotiations prior to Wapping, however, with management being able to demand massive redundancies and the removal of workplace demarcations (Littleton 1992; Tunstall 1996; Wintour 1989).

The *Financial Times* provides an excellent example of workplace relations in the national newspaper industry in the post-Wapping period (Marjoribanks, forthcoming). In July 1986, it announced major plans for the future development of the newspaper which would affect all those employed by the company. While the *Financial Times* had attempted previously to introduce new technology, as with other national newspapers the unions had been able to resist major reorganisation. With the events at Wapping, however, the company claimed that without workplace reorganisation the paper simply would not be able to remain competitive with the publications of News International. The plans announced by management at the *Financial Times* involved relocation of the paper, the introduction of computer technology, and a reduction in the production workforce from 592 to 206.

The plans for workplace reorganisation specified by management for the *Financial Times* were extensive and ambitious. The question that arose, however, was how to translate the plans into practice. One option was to follow the example of News International, and to impose the proposed developments on the work force. The management rejected this option. According to the Chief Executive, Frank Barlow:

> I do not intend to do a Wapping. I intend to do the very opposite of Wapping. I intend to do exactly what the print trade union leaders have always said is the proper way to achieve change. I intend to negotiate the introduction of a modern web-offset printing plant using the existing four printing and maintenance unions and drawing the work force from among our existing employees. I intend to do an anti-Wapping (1986:3).

Behind this goal of achieving reorganisation by negotiation was the non-negotiable requirement that the company achieve redundancies both quickly and on a large scale. As part of the objective to seek voluntary redundancies, the company set up an eighteen-month period to allow employees to plan their futures and negotiate with the company, and it offered generous terms for those who took redundancies. At all times, however, management made it clear that the redundancies outlined in the plan were absolutely necessary, clearly limiting the scope of possible negotiations.

In making these non-negotiable proposals for workplace reorganisation, the influence of Wapping was critical. While previous attempts to reorganise the workplace at the *Financial Times* had failed, events at Wapping were to open up a new range of options for management. A senior manager recalled:

> When [the dispute at Wapping] was more or less over, and the unions had realised what had happened, they were still reeling from the shock that they could not prevent newspapers from being printed and being delivered. So we said to our unions, you have two choices now. You either go the way that we want you to go or we'll do a Wapping. We don't care. Do what you want, you know. So they said, well, we don't really want another Wapping, we'll

negotiate with you – which we did. But that enabled us then to look at technology quietly and carefully at our own pace, and say what technology we want to use for our business for the future.

That the events at Wapping had a significant influence on relations between management and employees is reflected in the comments of a senior member of management who recalled an immediate modification in the atmosphere and practice of workplace relations:

> The night we announced [the development plans for the *Financial Times*], we announced to all our workforce, even those not affected by the plans, a series of big presentations, and we sent videos to all the homes, but I remember clearly the night in this big hall in the city we told all the print workers what had happened, and a week before they would have all walked out. They wouldn't have been out of a job; they would have walked out of the hall. But they sat there, and they went back to work, and they produced a newspaper, and they would never have done that months before. Wapping traumatised; Wapping helped us to get it through because it traumatised the workforce throughout Fleet Street.

Wapping was crucial for management, as it illustrated that workplace developments could occur without the agreement of the unions. As has been noted, however, management at the *Financial Times* did not necessarily consider that pursuing the 'Wapping model' was the best way of achieving change in their organisation. Rather, they sought to at least keep the workforce informed of proposed developments, while using threats of Wapping to secure the cooperation of the unions and the workforce. According to a senior manager, the negotiations at the *Financial Times* were carried out with employee participation, with the situation of News International being used as an example of a possible alternative approach. He suggested 'people have to understand that your preference is to do things in as a humane, sensitive way as possible, but if that doesn't work then all bets are off. They [the workforce] understood that because they had the example of Wapping'. While change would be negotiated, it was to be negotiated on terms decided by a management that was prepared to use the tactics of News International if necessary. The compliance of the unions with the demands of management indicates that they perceived the threat of derecognition as real, and chose a course of action which at least meant they would be recognised at the new worksite.

A significant consequence of the developments in the 1980s at the *Financial Times* has been a fundamental reorganisation of relations in the workplace involving management, unions and individual employees. In contrast to the case of News International, where trade unions have been derecognised, at the *Financial Times* the unions are still recognised. However, the role of unions in workplace negotiations is much reduced. One senior journalist argued, for example, that while there remains a high degree of unionisation at the *Financial Times*, the situation was very different post-Wapping. While applauding the management for seeking to negotiate with the workforce in a number of areas:

> [i]t has to be said that [management has] taken advantage of the general climate of recent years, and getting them to negotiate in a serious and worthwhile way has become much more difficult. Yes, we still have negotiations. Yes, that is enormously important. I don't want to underestimate that. It's important because if you lose it, it's hard to get it back again. It's vitally important because it enables us to represent individuals, and some of the most important work that the union does is for individuals. But there is no doubt that the annual negotiations have – I wouldn't call them a sham – but they are certainly a shadow of the way that negotiations ought to be. Because the management knows, in its eyes it would say it's doing the unions a favour by continuing to recognise them, when other managements haven't.

As with other national newspapers, the *Financial Times* was able to take advantage of a new workplace environment created by the actions of News International to introduce new technology and workplace reorganisation. While the company continued to recognise the newspaper unions, the combination of Wapping and an institutional and societal context that had made Wapping possible meant that workplace relations were organised in a profoundly different manner.

Conclusion

The result of the move of News International from Fleet Street to Wapping has been the establishment of a union-free workplace in which workers are hired on individual contracts, and in which the right of management to manage without constraint from the workforce is enforced. The present analysis indicates that the relationship between the introduction of new technology and workplace reform is influenced by a complex set of interacting factors such as government policy, legislation, the coercive forces of the state, and the links between employers and employees, as well as the prevailing economic and social conditions. In the particular case of News International, the study reveals that workplace reforms were related intimately to an agenda pursued by government and the employer that included weakening the influence of trade unions in the newspaper industry. The investigation indicates that the eventual practices adopted in the workplace resulted, in part, from divisions in the union movement, suggesting that theoretical approaches need to explore variations in workers' involvement in workplace reorganisation. The conflicting practices of the traditional print unions and of the EETPU were vital to the eventual outcome of workplace reorganisation at Wapping. Also, the analysis of News International suggests that the nature of workplace reforms were related to the particular goals and attitudes of the Murdoch management.

The relationship between technological innovation and workplace reorganisation at the *Financial Times* offers an interesting comparison to the case of News International's newspaper holdings in Britain. Operating in a similar political, economic and social climate as the management at News International, developments at the *Financial Times* occurred without the intense and hostile confrontations that occurred at News International. Unions remain recognised at the 'new' *Financial Times*, 'negotiations' between the parties in the workplace occur, sophisticated new technology has been introduced, and significant workplace reorganisation has proceeded. The vital difference between the cases, however, is that the developments at Wapping had occurred. Those events had so shocked and drained the union movement, that management's plans – which may well have been rejected out of hand by the unions only months before – were accepted by the union movement, even though it meant the virtual decimation of the print labour force.

At the same time, the ideological differences between management at News International and the *Financial Times* should be recognised. At News International, the unions have been derecognised. In contrast, members of both management and the workforce at the *Financial Times* recognise that management is at least somewhat sympathetic to the unions. On this point, however, the developments at Wapping must be remembered. The outcomes of those events, shaped by the social and institutional conditions of the period, allowed the management at the *Financial Times* to appear 'liberal' in their dealings with the workforce. The alternative of 'Fortress Wapping', the continuation of issues of unemployment and technological advances resulting in fewer labour power requirements, have meant that management can afford to recognise the unions, and allow for modes of workplace representation by unions, while dominating both technological developments and processes of workplace reorganisation.

In general, the analysis in this chapter supports the relational model of workplace reorganisation by showing that the relationship between technological innovation and reform in workplace relations is influenced by a set of interacting forces that include the government, legislation, the state, and the relation between employer and employees operating within national and international social and institutional contexts. Furthermore, the study of the *Financial Times* reveals that workplace reorganisation in one firm can be dependent on what has occurred in another firm within the industry. That is, perceptions by management of the opportunities afforded by events in another firm and perceptions by workers of the possible costs to their work security and conditions if these events were repeated, can influence eventual workplace outcomes. Therefore, theoretical approaches that examine the nature of the relationships between technological innovation and workplace reform need to take into account such factors as the possible impact of the historical moment in which workplace reorganisation occurs and the perceptions that actors have of each other's power over the processes. The analyses in this chapter indicate that workplace reorganisation at one site can influence greatly the processes of workplace reorganisation at another setting within the same country. In the following chapter, I examine how the experiences of management and workers at Wapping differed to those of the people involved in newspapers in the Australian context.

CHAPTER 6

The Adelaide Advertiser: Wapping South?

Technological innovation and workplace reorganisation were achieved at the newspapers of News International in Britain in the mid-1980s only after a bitter and violent dispute between the company and newspaper trade unions, resulting in the establishment of a non-union workplace at Wapping in the Docklands of London. While not making use of the most recent technology, the shift in production technology from hot metal to computers at Wapping was the catalyst for overwhelming workplace reorganisation in the national newspaper industry in England. Soon after the Wapping dispute, News Limited in Australia commenced a major program of investment in technology and plant at its newspapers, including the Adelaide *Advertiser*. While these investments and accompanying processes of workplace reorganisation gave rise to some conflict between unions and management, and while the new production plant in Adelaide was even referred to for a period as 'Wapping South', the outcome in Adelaide has been a workplace where workers are still represented by unions, and where unions are still involved in negotiations with management over issues relating to technology and organisation of the workplace.

The *Advertiser* is the only daily newspaper in Adelaide, the capital city of South Australia, with a population of approximately one million people. Until 1987, the *Advertiser* newspaper was owned by the Advertiser Group which was formed in the early 1960s from diversification involving the *Advertiser* and other organisations into a group of companies in the printing and packaging industries. This diversification was accompanied by a reorientation by the management of the *Advertiser* 'away from an editorial perspective and towards greater concern with market development' (MacIntosh 1984:33). MacIntosh suggests that the process of diversification and development was 'evidence of an entrepreneurial

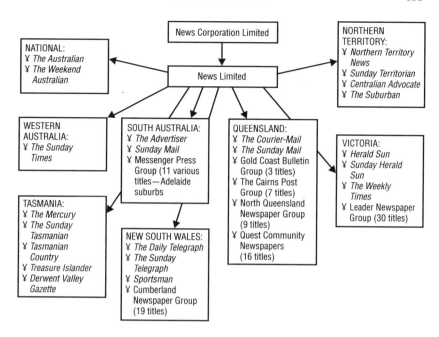

Figure 6.1 Australian Newspaper Holdings of News Corporation Limited 1998

approach consistent with a degree of innovation and risk-taking in technological development' (1984:33). Ownership and control of the *Advertiser* passed to News Corporation Limited in 1987. In its *Annual Report* for 1986, the management of the *Advertiser* reported that by taking the decision to accept the offer of News Corporation, 'the *Advertiser* has been able to participate in the largest media group in the western world' (Advertiser Newspapers Limited 1986:20). The *Advertiser* was also now part of what was to become the largest print media organisation in Australia. As Figure 6.1 illustrates, by 1998, News Limited had extensive newspaper interests in all major Australian cities, and in Adelaide, Brisbane, Darwin and Hobart it possessed a monopoly of newspaper ownership.

While the focus of this chapter is on the *Advertiser*, its connections with other News Corporation newspapers, both inside Australia and overseas, are also considered to allow for further examination of the practice of News Corporation locally, nationally and globally.

New Technologies at the *Advertiser*: A Tradition of Innovation and Implementation

At the *Advertiser*, computer-based technological innovation in the production of newspapers has been occurring since the 1960s. In early

1967, the *Advertiser* internal staff magazine *Pi* reported that 'a computer is now helping to set type for the *Advertiser* ... giving us, in the words of our production people, much greater flexibility' (*Pi* 1967a:1). This new system, which involved computerised setting of print, was faster than the preceding manual linotype and intertype setting system. The novelty of the system in 1967 was apparent from the language used to describe computer programming: 'it [the computer] has been given carefully worked out instructions about setting which have been fed into its electronic memory' (*Pi* 1967a:1). Twelve trained operators worked on this computer typesetting system, which was the first to be used in an Australian newspaper (*Graphix* 1981b). Further computer-related innovations in the workplace occurred in 1967 with the company reporting that 'almost a new language is spoken [in the press room]', while the 'complexity and sophistication' of the machinery in the press room required the appointment of an assistant chief engineer and of a press room superintendent (*Pi* 1967b:6). By early 1974, the *Advertiser* had installed a computerised telephone system to assist the Classified Advertisements Department, with thirty-three positions becoming available (*Pi* 1974). Investment in technology clearly affected many areas of the publication of the newspaper.

A fundamental shift in the production process at the *Advertiser* occurred in 1977 with the introduction of the electronic newspaper system (ENS). The system brought electronic processing techniques to the *Advertiser*, and signalled the imminent end of hot metal production. Speaking at the time of its introduction, the Advertiser Group Managing Director P.J. Owens said 'The system will consist of editorial, classified and display matter being entered into the system via [video display] terminals located in the editorial, classified and production departments and being stored, corrected, revised and output from a computer base' (*Pi* 1977b:2). Owens provided a number of reasons for introducing the new system, including the need to be able to handle increasing paper sizes and advertising volumes, a decline in available labour creating a need for greater mechanisation, the challenges of competition from other media, and a need to stop high cost escalation (*Pi* 1977a). While many print workers would lose their jobs, for journalists the introduction of the new system represented a substantial opportunity. According to one journalist at the *Advertiser*:

[The introduction of the ENS] was a very exciting period, it was really great, because from a journalist's point of view the system gave you so much more flexibility, particularly to a sub-editor. For example, in hot metal, if you wrote a heading and it was set in metal but it didn't quite fit, it would have to come back. The piece of paper would come back with a note saying 'you've missed a half a letter', whereas with the new technology you can see that on a screen instantly, and you can change the size of your type by half or quarter of a point . . . You could do much more in the way of designing pages, making your stories fit, can turn corners. We've been given rubber print in effect.

The introduction of computer-based technology at the *Advertiser* had substantially different effects on the labour force depending on the location of the particular job in the workplace. Retraining programs were established, however, with the promise that '[n]o-one will be made redundant as a result of the introduction of the ENS system' (*Pi* 1977a:15). Print workers were assured further that '[a]t all times close liaison will be maintained with your unions in relation to new job classifications, pay rates, aptitudes, training programs and so on' (*Pi* 1977a: 15; *Pi* 1979; Advertiser Newspapers Limited 1979). It was intended that employees of the company who so desired would be retained in employment, although it was also recognised by Owens that 'as years go by this newspaper will use fewer people on many aspects of its production' (*Pi* 1977a:15).

The introduction of the ENS was completed on the night of Friday 8 May 1981 when the final Page 1 of the *Advertiser* was produced by hot metal typesetting methods (PTJ 1981; *Pi* 1981). It was announced that '*The Advertiser* is the first newspaper in the world to convert entirely to a system which, it is believed, in the next decade will revolutionise newspaper offices everywhere' (*Pi* 1981:6; *Graphix* 1981b). According to the General Manager of Advertiser Newspapers Limited, Brian Sallis, the investment in the ENS constituted 'one of the most important, and most exciting, production developments in *The Advertiser*'s history' (*Pi* 1977c:6–7; Advertiser Newspapers Limited 1978).

These momentous developments involved major workplace reorganisation. According to Brian Sallis, the '[i]ntroduction of electronic processing techniques to the editorial, advertising, production and commercial aspects of *The Advertiser* required a complete change in our thinking and knowledge of publication techniques for newspapers' (*Insider* 1993d:6, see also 1993a, 1993b, 1993c). Commenting on this same process, Commissioner Donaldson in the Australian Conciliation and Arbitration Commission reported that 'the smooth transition to this new technology was accomplished by protracted and detailed consultation between management, employees, and the PKIU' (Donaldson decision 1984:463). While the changes associated with the ENS were major, significant processes of innovation and reorganisation at the *Advertiser* were to continue in the 1980s and into the 1990s.

The *Advertiser* announced plans in September 1985 to establish a new publishing and printing centre at Mile End, a suburb of Adelaide close to the pre-existing publishing centre in the city. Brian Sallis, by

1985 the Group Managing Director of Advertiser Newspapers Limited, was reported as saying, 'it would be one of the most modern publishing centres in the world' and would involve the introduction of more extensive colour facilities, resulting in '[b]etter press performances, both in speed and quality of product' (Hammond 1985:1; Advertiser Newspapers Limited 1985). It would also allow the *Advertiser* to increase its production capacity beyond what was possible in the increasingly cramped city location, while also introducing on-line automatic inserting equipment (*Pi* 1985b). On announcing the development plans, Sallis claimed, 'I personally think that this and the ENS have been the most significant steps forward in the last 50 years for Advertiser Newspapers Limited, and I am very excited about it' (*Pi* 1985b:9). The project was to proceed through a number of stages. First, a paper reel store would be built, followed by the construction of a press hall and the installation of a new multi-colour press. Once these investments had taken place, 'it is intended to update, renovate and install modern colour units on some of the old presses currently in the city. The sections of the *Advertiser* to transfer from the city will be the plate, publishing and press rooms plus support staff. These will include engineers, electricians, computer technicians and some cleaners. Editorial and composing sections will remain in the city' (Mortimer 1985:145). According to the Printing and Kindred Industries Union, 'Management of the *Advertiser* spoke to all employees at the *Advertiser* and Messenger Press and explained a concept model showing the proposed printing centre at Mile End' (Mortimer 1985: 145). This process of notifying the workforce departed substantially from the practice at Wapping where unions were not informed of the plans for workplace reorganisation.

No deadline was established for the costly development, but work on the new plant began in earnest with the takeover of the *Advertiser* by News Corporation in 1987. In March 1988, News Corporation placed 'what was believed to be the world's biggest ever press order with the MAN-Roland company, worth more than A$700 million' (*Graphix* 1988b:5). Such was the size of the order that MAN-Roland renamed the presses MAN-Newsman 40 to acknowledge the investment by News Corporation in their product. Later in the same year News Corporation announced that a A$126 million printing plant was to be built for the *Advertiser* in Adelaide as the first of four such News Corporation plants around Australia. According to the Australian printing industry publication *Graphix*, 'The *Advertiser* plant, which will be located at Mile End, will incorporate three of the four most modern newspaper presses in the world. The four-colour presses are said to be the fastest in the world, producing up to 80,000 newspapers each hour' (*Graphix* 1988b:5).

In addition to publishing the *Advertiser*, the presses would publish other Adelaide-based News Corporation titles including the *Sunday Mail*, the *News*, suburban newspapers of the Messenger Press, and local copies of the national daily the *Australian*. Rupert Murdoch claimed that the new presses, which would be part of the most modern printing plant in Australia, would allow readers to enjoy better quality print and later breaking news (*Graphix* 1990a; *Pi* 1990). The Premier of South Australia, John Bannon, announced that the investment in the plant by the company was 'a feather in the cap' for South Australian industry (*Pi* 1990:5–6). To further the development of the new printing centre, News Limited signed a A$12 million agreement with a United States manufacturer to supply an automatic guided handling system for newspaper production (*Graphix* 1990b:3). The company also invested in technology for the editorial and composing functions of the newspapers. In 1993 News Limited signed multi-million dollar contracts for Cybergraphic advertising systems for a number of its newspapers. Among the newspapers, '*The Advertiser* has contracted for a 152-terminal advertising system, classified pagination, display ad makeup and an upgrade to its existing 170-terminal editorial system' (*Graphix* 1993a:3). On completion of the Mile End plant, which ultimately cost A$148 million – made up of A$100 million for equipment, A$42 million for the building, and A$6 million for an adjoining paper storage centre – production of the newspapers would occur as a flow-through operation in which complete newspaper pages prepared in the city would be transmitted to the Mile End plant by facsimile, where the material would be fed in at one end of the production system and come out the other as a complete product (*Pi* 1989). Commenting on these developments, a senior manager at the *Advertiser* stated that 'the beauty of it here is that we had the situation at the time of starting from a bare patch of ground so people were able to watch it develop'. He went on, 'The whole place reeked of quality, of brand new equipment. There's not a second hand nut or bolt in the place.' As an integral part of A$1 billion to be invested by News Limited around Australia in its newspaper interests, the *Advertiser* was to become part of the most expansive development in the history of Australian newspapers.

After the completion of the Mile End plant, reporter Michael Atkin of *Graphix* reported that 'News Limited's *Advertiser* and *Sunday Mail* in Adelaide are produced in a plant which is a shining example to other newspaper proprietors in Australia' (Atkin 1992a:19). Atkin reported that 'we were given a tour of the plant, seeing first-hand the technology that News Limited were taking with them into the next century. The press and publishing room is fully computerised and robotized. Ferag and others are responsible for the veritable jungle of conveyors and transportation belts that filled the publishing room' (Atkin 1992a:19). In addition to the 'high-tech, low-people-cost plant, built from scratch at Mile End close to the city' which Atkin labelled 'the present and future of newspaper production all rolled into one', the company claimed that 'the training program put together to bring the staff up to speed with all the technology is ... itself an exportable product' (Atkin 1992a:19). While the immediate beneficiaries of such an export would be the other Australian-based plants of News Limited, it was anticipated that globally-based facilities of News Corporation might also benefit from the experiences in Adelaide. The possibility of exporting the developments made at Adelaide was also alluded to by Rupert Murdoch who, on visiting the newly finished plant in 1991, commented that it was a model in world terms for atmosphere and working conditions (*Insider* 1991b). The potential influence of these developments on other newspaper companies was also apparent from the statement of a representative of the Melbourne-based newspaper consortium, Australian Independent Newspapers, who claimed:

> We have to remain competitive with organisations such as News [Limited] and it is clearly setting a fairly fast track on technological development and looking also for efficiencies. We would have to do that to compete ... We do have a great belief in printing and publishing in the 1990s and we base a lot of that on the improved technology that can be brought to bear in terms of productivity, in terms of quality and in terms of colour. So we would like to do it (install new technology) as early as is financially prudent and relate it to total market conditions and particularly the strategies adopted by News [Limited] (AHRSC 1992:144).

While these developments were occurring at the *Advertiser*, News Limited was implementing similar improvements in its plants in Melbourne, Sydney and Brisbane. In late 1992 and early 1993, work was completed on a 'A$350 million plant at Westgate Park close to the centre of Melbourne' (Atkin 1992b:7). The Melbourne plant is the biggest of the four plants with six presses, compared to five in Sydney, four in Brisbane, and three in Adelaide. Commenting on the Melbourne plant, which was designed by the same organisation that had worked on the development in Adelaide, Michael Atkin reported that this link between the plants 'gives a feeling of continuity between the associated manufacturing plants, offering the back-up and training benefits obviously in the mind of the architects of the overall News Limited program' (Atkin 1992b:7). In 1998 it was announced that three state-of-the-art presses would also be introduced at the News Limited printing complex at Canning Vale in Perth (News Corporation Limited 1998).

With the completion of the Chullora plant of News Limited in Sydney earlier in 1994 at a cost of A$360 million, it was reported that 'News [Limited] is nearing the completion of its investment of more than A$1 billion in a modern national printing and publishing network' (Kohn 1994:12). Further investment in the News Limited plant at Chullora in Sydney was reported in August 1994 when the production system was upgraded by introducing the Cybergraphic Editorial System. Comprising 430 editorial work stations, and said to be 'one of the largest and most advanced computerised editorial production facilities in the world ... the new system will incorporate fully electronic on-screen pagination, interfaces to new image and text libraries, pagination capabilities for workstations at remote sites, and ability to handle full colour production. It will also integrate booking and scanning of display advertisements so that entirely completed pages can be produced' (*Graphix* 1994:7). Given the intention of News Limited to develop an integrated nation-wide production network, this system was to be introduced into its other Australian plants. In discussing the new plant, Garry Walsh, Deputy Plant Manager at Chullora, summarised the News Limited philosophy: 'When you start from scratch, you have a once-in-a-lifetime opportunity to do it right. We've allocated lots of growing room here' (Kohn 1994:12).

The 'end result' of this massive Australia-wide project of technological innovation for News Corporation was the introduction of full editorial pagination at all of its Australian newspapers. Identified as 'the biggest advance in newspaper production since hot metal gave way to the first typesetters' (Mitchell 1997:2), the electronic pagination process allows an entire page comprising text, graphics, pictures, advertisements and all rules and page numbering to be created and brought together on a single computer screen (*Insider* 1996:2). The *Advertiser* was the final newspaper in News Corporation's stable of Australian newspapers to introduce full electronic pagination in August 1997, allowing it to build on the experiences of newspapers in Sydney, Melbourne and Brisbane. According to News Limited's Group Computer Services Manager, the completion of this project made the metropolitan newspapers of News Limited unique globally because they would all now be fully paginated and output either digitally or to film (Howarth 1997).

As a result of these investments in production technologies, the 1980s and 1990s involved negotiation and struggle over the influence of the new technology on the workforce, with critical issues being job demarcations, pay rates and employment levels. These issues were similar to those that had confronted the newspaper industry in Britain in the same period. The different societal and institutional Australian context, as well as the history of workplace relations at newspapers such as the *Advertiser* meant, however, that workplace reorganisation came about in a very different manner.

Technological Innovation and Workplace Reorganisation at the *Advertiser*: Negotiation or Coercion?

Formal Regulation of Workplace Relations

Workplace relations and negotiations at the *Advertiser* throughout the 1980s and the 1990s were governed by three major federal awards. Print workers were covered by the Newspaper Printing Agreement, clerical workers by the related Newspaper Printing (Clerical and Associated Officers) Award, and journalists by the Journalists' (Metropolitan Daily Newspapers) Award. During the 1980s and 1990s, these awards were renegotiated to accommodate issues such as the introduction of technology. They were also supplemented by new awards such as the Metropolitan Daily Newspapers Redundancy Award 1996 and the Journalists' (News Limited-Metropolitan Daily Newspapers) Certified Agreement 1996.

The Newspaper Printing Agreement 1979 covered print and production workers at the *Advertiser*, at an Adelaide afternoon newspaper entitled the *News*, and at a newspaper in Hobart. Certification of the Agreement by the Australian Conciliation and Arbitration Commission was acknowledged by the PKIU in January 1980 as 'the beginning of a new era in the production areas of the Metropolitan Daily Newspapers in Adelaide and Hobart', as it introduced improved conditions and regulated technological innovation in the workplace (PTJ 1980:5). In addition to regulating wages, job demarcations and other conditions of work, the Agreement contained clauses which recognised that new technology which might require adjustments in conditions of employment was being used in newspaper production. To this end, the Agreement made provision for demarcation between compositors and journalists in the use of equipment, as well as providing for retraining and reallocating workers affected by computer technology. Importantly, the Agreement also provided that there were to be no redundancies directly associated with the introduction of the ENS. The union was also to be notified three months in advance of the introduction of any new technology affecting the work methods of its members. That is, the Agreement provided a clear framework for the introduction of new technology in response to the concerns of both management and the organised workforce.

An important insertion into the award was made in 1989 relating to structural efficiency. In order to comply with the 1989 National Wage Case decision handed down by the Industrial Relations Commission, it was ruled that employees were to perform a wider range of duties, and that a single trade stream was to be introduced for the pre-press area. The goal was to create a job structure in which functions were broadly-based and generic.

In 1996, a new award was made between the PKIU, by now part of the Automotive, Food, Metals, Engineering, Printing and Kindred Industries Union (AFMEPKIU), and News Limited, entitled the Metropolitan Daily Newspapers Redundancy Award. This award recognised that continuous change in work methods related to technological innovation would be part of the newspaper industry for the foreseeable future, requiring a separate award to cover the impact on employee numbers. Again, News Limited was required to give at least three months' notice of the possibility of technology redundancies, and provisions were made for voluntary redundancies and for redeployment of affected individuals within the organisation. Provisions were also made in the award for those who became redundant to receive financial counselling and time to seek alternative employment, and to be given a minimum of four weeks' notice of termination of employment on the basis of technology.

Workplace terms and conditions for journalists at the *Advertiser* were covered by the federal Journalists' (Metropolitan Daily Newspapers) Award 1974, which was varied in 1982 to become the Journalists' (Metropolitan Daily Newspapers) Award 1982, and underwent a number of further variations in the 1980s and 1990s. As with the Newspaper Printing Award, the Journalists' Award covered wages and conditions for journalists working for a number of metropolitan daily newspapers, including the *Advertiser*. Training was to be provided for journalists using visual display terminals (VDTs), and provision was also made requiring employers to consult with the AJA and its members affected by changes of technology deemed to have a significant effect on the work of AJA members.

In 1996, the Journalists' (News Limited-Metropolitan Daily Newspapers) Certified Agreement was introduced which was binding on the *Advertiser* and other News Limited newspapers. The Agreement was introduced to create an efficient and skilled workplace at organisations such as the *Advertiser* by managing the implementation of replacement and new pre-press computer systems, award restructuring, redundancies and the elimination of unnecessary demarcation barriers. In particular, demarcations on the use of technology were to be eliminated as new systems were introduced. As part of this process, provision was made for employees who were redeployed from composing rooms to the editorial department as a result of technological innovation relating to wages and conditions of employment. In this way, the Agreement was introducing a new regulatory framework for workplace restructuring related to technological innovations that made previous demarcations potentially redundant.

These awards and agreements provided the formal legal foundation for workplace relations at the *Advertiser* in the 1980s and 1990s. The more recent developments, in particular, recognised that this was a period of profound reorganisation for newspapers, related to technological innovation, which required a framework to allow workplace transformation to occur smoothly and in an acceptable manner for all parties. Negotiations and disputes arising in this period illustrated, however, that the substantive content of the legal framework was affected by local workplace practice.

Workplace Relations on the Shopfloor:
Cooperative Paternalism

In a study of the *Advertiser* from the 1980s, MacIntosh identified the relationship between management and workers as paternalistic. For MacIntosh, an important element of paternalism was 'the assumption that the employees' interests and aspirations are in harmony with those of management' (1984:34). Other elements of paternalistic personnel practices at the *Advertiser* included the employment of family members, employment security through the provision of 'no redundancy' agreements, and an emphasis on communication with the staff. MacIntosh also identified a coercive element to the relationship in that those who questioned management practices could be labelled as troublemakers, and be subjected to counselling. The paternalistic relationship assumed that employees would be faithful and loyal to decisions and guidelines set by management. According to MacIntosh, although elements of coercion existed, the resultant relationship between management and labour at the *Advertiser* was founded on trust.

Previous studies of technological innovation governed by the Newspaper Printing Agreement at the *Advertiser* in the early 1980s have emphasised the relatively peaceful and cooperative means by which innovation was negotiated and executed. Analysing the introduction of the ENS, Patrickson observed that '[a] basically cooperative strategy, which excluded the possibility of any worker redundancy, contributed to a comparatively trouble-free introduction of the system' (1986:3). According to Patrickson, 'the overwhelming thrust of the technology was such that', rather than opposition to the technology, the 'efforts of printing employees became directed at making the system run more smoothly and union effort was directed towards the best level of consequent benefits for their members' (1986:3). MacIntosh also identified this cooperative interaction between management and unions, and suggested that the negotiations over the introduction of ENS 'could not have occurred [in such a manner] without a climate of trust between union and management' (1984:37).

As an example of cooperation between management and unions, the company agreed to the 'no-redundancy' clause in the Newspaper Printing Agreement, aware that this would lead to higher costs in terms of retraining and payments for redundancy where necessary, but also aware that labour costs would decrease with attrition in the workforce. At the same time, the PKIU agreed that the company be allowed to hire casual and contract labour for a period when the technology was being introduced. The union also supported the plan of the company to restructure composing classifications into fewer categories (MacIntosh 1984). None of these agreements was necessitated by the ENS, and in each instance could be considered to be contrary to the interests of one of the parties. Casual and contract workers are often not union members, and can be difficult to organise into unions. The hiring of such workers may not be in the interests of unions, or of unionised workers. That the PKIU agreed to this measure may be taken as an indication of the influence of a combination of factors, including an acceptance that technology represented progress (Patrickson 1986), and that previously developed relations between labour and management were founded on trust, meaning that company promises of no redundancy would be honoured. The workplace restructuring agreements entered into can also be taken to represent the power invested in management by the potential labour-saving dimensions of the new technology.

Training, Retraining and Demarcations

While there were examples of cooperation between management and unions in the process of technological innovation and workplace reorganisation, cooperation did not necessarily mean the same thing for the two parties. An example of these differing perceptions and experiences of cooperation is related to the establishment of extensive training and retraining programs for employees associated with the introduction of new technology. As part of retraining for the ENS, nine members of the production staff were sent overseas to undergo training on presses, while extensive lectures for all staff members and employees were conducted (*Pi* 1977c). In addition, stereotypers made redundant by ENS were not retrenched, but were dispersed to acquire skills in areas including pre-press, publishing and composing. Ten stereotypers were posted to the composing room, and were given the opportunity to become fully-qualified trades persons through courses and training. Eight became qualified, one withdrew for health reasons, and another took up full-time union duties with the PKIU (*Pi* 1985a). This concern with training and retraining has continued into the 1990s, with the emphasis now being placed on management as well as employee training. The newspaper has implemented extensive training programs for all existing and new members of staff and management, with the aim of giving 'to all members of staff and new employees a full overview of the skills required and of company objectives' (*Insider* 1994:8). Staff members from all the departments of the newspaper have attended courses on issues ranging from customer service to management, development and supervision (Monk 1996:5). The newspaper has also hired management consultants, who provide training to implement total quality management, by which they mean workplace organisation based on consultation between management and the staff, with training being an important element of fostering and maintaining such communication, most recently in relation to electronic pagination (*Insider* 1991f; Klima 1997; Mitchell, Hart and Owen-Brown 1996).

For management, these training and retraining programs represented a triumph of coordination between management and workers. Speaking about the introduction of ENS, for example, a management representative reflected:

> We went to great lengths to make sure that our staff that wished to be involved in it [in the introduction of ENS] were involved in it, were properly trained, and as a part of that we made sure that the unions and the chapel were all part of the planning; they were all part of the training. There were many, many discussions.

Extensive training programs were also required in the late 1980s and into the 1990s for the relocation of the production plant to Mile End, as noted by another management representative:

> One of the cornerstones of the exercise was that we knew that we had an enormous retraining requirement ... The process involved three weeks of theory and practical training [of print workers] at a training school set up in Melbourne by News Limited, and then the people came back and were progressively trained in the various elements of the new presses, and those elements of the operation were broken into a number of modules. People were taken through these modules however many times they needed to be taken through them ... It's a very organised process, there's a committee down there that assesses the needs of people, that monitors it, and goes

through the whole process. We have a full-time trainer in the press area still at Mile End.

This latter person indicated that similar steps had been taken in other areas of production and distribution, while managers had also been trained to understand the new technology, resulting in 'a total retraining process'. Again, it was claimed that these programs involved all parts of the organisation in its development:

> A lot of our employees were themselves a part of the development of the system, and we tried to develop a system that was designed for the *Advertiser* . . . So there was a lot of work and quite a number of people who probably at the time would never have contemplated themselves being involved in the development of a newspaper production system suddenly found themselves working there with these hardcore programmers developing a newspaper production system.

Union members, however, had a different interpretation of these programs. One printer, for example, argued '[w]e've had to fight for everything. Management won't give anything without making it a fight. Retraining, for example. They always want to bring in new people'. For union members, while training and retraining programs were preferable to compulsory redundancy, they were also interpreted as an attempt by management to disguise the real impact of redundancies induced by the introduction of technology, or as a means of introducing lower paid workers into the workplace.

Retraining itself was also not necessarily considered to be beneficial to the workforce, with retraining for positions that removed preexisting demarcations considered to be particularly threatening. This threat continued into the 1990s with compositors at the *Advertiser* engaged in a struggle to maintain 'manual' cutting and pasting, and resisting the introduction of computerised pagination which would result in the loss of many jobs. One compositor claimed: 'We won't let them do pagination; we want to keep the jobs separate', while going on to say:

> Demarcation issues are important. We don't have pagination yet because we have demarcations between different areas; between editorial, advertising, and stories; keep stories separate. Management wants to get rid of the reading room. We demand to keep those jobs from the point of view of quality control. Other papers have got rid of readers and their quality is terrible. We win quality control prizes, and we want to keep it that way.

Despite the resistance of compositors, pagination was fully introduced in 1997, and as evidenced in the recent awards and agreements of both the PKIU and the Media, Entertainment and Arts Alliance (MEAA), demarcations have largely disappeared. As part of the process of introducing electronic pagination, News Limited has replaced composing rooms at its newspapers around Australia with edition control centres. At the *Advertiser*, edition control is located within the editorial department, and is run by production editors (Howarth 1998). Once journalists have entered their stories on terminals, layout sub-editors design the pages of stories and photos. The Edition Control Centre does a final check of the pages, which are then sent to the plate room (Howarth 1997; Klima 1997). Electronic pagination has clearly had a major impact on the organisation of the newspaper workplace. It is important to recognise, however, that the unions have been involved in this process of restructuring, in complete contrast to the English experience. Through such involvement the unions have been able, for example, to win agreement from News Limited that 'printers could perform computer-based tasks on an editorial pagination system operated primarily by journalists' (Hogben 1994:17).

Demarcation issues have also arisen within the production section of the newspaper. In 1991 it was reported in the company newsletter that in the move to Mile End, the print division had been 'more extensively and structurally altered in the past couple of months than at any other time in the company's history. Just about every employee in the Press Room, Storeroom, Publishing and Engineering has changed responsi-bilities in a geographically different environment' (*Insider* 1991e:2). The result was a closer coordination between the various sections, with training and workplace structures emphasising greater flexibility among the staff. While management was concerned to promote flexibility and fluidity between tasks, such restructuring also resulted in a reduction in labour requirements among printing and production workers.

Technology and Job Loss

Demarcation issues in turn give rise to concerns over job loss associated with technology and workplace reorganisation. While the shift to ENS resulted in a reduction in the production staff of some 20 per cent, the move to Mile End in 1991 was accompanied by the loss of about eighty members of the production department staff, who took the redundancy package offered by the company prior to the transfer. The company emphasised that those taking the redundancy package did so on the basis of individual choice (*Insider* 1991a, 1991c, 1991d). Major reductions have also occurred in the compositing staff. In 1988 after the takeover of the *Advertiser* by News Limited, there were 267 members of the composing staff, and that number had been reduced to 64 by

1995. This reduction was associated with the labour-saving capacity of computer technology, and has also been attributed by the company to other factors, including the closure of the *News* (an evening newspaper), the loss of a publishing contract with the *Messenger* newspaper, and outside advertising production (*Insider* 1995b). The last development, in particular, can also be linked to changes in technology, as agencies can produce high-quality advertisements for companies which previously relied on newspapers to fulfil that function. As a compositor at the *Advertiser* commented, 'The big threats at the moment are coming from [advertising] agencies. Computer technology has made it easier for them to put together a top quality advert' (*Insider* 1992a:2). At the same time, this compositor was critical of the newspaper itself for contracting out the job of colour-separating, and not giving it to in-house compositors. Such contracting out of jobs is a further threat to the positions available within the production process, and is linked to the search by the company for further ways to reduce costs, and to promote flexibility in the workplace.

Technology and Skill Requirements

Negotiation does not occur, however, only at the point at which technology is introduced into the workplace. Adaptation to the requirements of new technology in terms of skills requirements, for example, takes time. Over such adjustment periods the relations between management and labour may either improve, as both parties adjust to the new conditions, or deteriorate, as adjustment difficulties continue. Patrickson observed that after the introduction of ENS, employees at the *Advertiser* 'perceive they have lost considerable ground in their negotiating power within the organisation which many associate with the loss of exclusiveness' (1986:6). As computer technology is introduced to the workplace, so the craft nature of the work of the printer becomes replaced by the different skills of a computer operator. One worker commented, 'We're pretty powerless really. We could stop three to four hours and the paper would still go out', while another worker claimed 'There's not much we can do. We can't threaten industrial action because they can get around it' (Patrickson 1986:7). Patrickson suggested that the new technology makes skills easier to learn, and workers become replaceable: 'Computerised newspaper production has permitted management to replace manual skills, coordinate operations, and to restructure the production process to lessen its dependence on the input of the skilled compositor' (1986:8). While print workers and their unions can still enter negotiations, their position of entry into negotiations has been weakened by the introduction

of technology that gives management more flexibility in hiring practices and in labour requirements. Such a situation has been
exacerbated for the workforce by the contracting out of various jobs.
In early 1995, seventy positions in the pre-press production area were
lost when Messenger newspapers ended its printing partnership with
the *Advertiser*, and entered an agreement with a smaller press (*Insider*
1995a). With the availability of production tools more widespread,
related to the development of desktop publishing, the sources of job
loss threat for the traditional employees of newspapers are spreading.

In addition, there exist different understandings of the potential
impact of new technology on work requirements, and these understandings influence the negotiation process. A management
representative claimed, for example, that 'from 1979 through to 1990,
in that decade the changes in demarcation were probably not significant'. He continued:

> Basically it meant that you went from the hot metal process to computer
> generation – but it tended to be a replication of functions . . . so what you
> had was the computer replicating the hot metal process without the indus-
> trially heavy function. About the only thing that possibly came out of this was
> the fact that you had the circumstance where the journalist could then sit at
> the computer screen and input his material, rather than typing it on a type-
> writer, and then sending it to someone else to put into the system.

Finally, he argued that 'the change was more a situation of people
having to be updated to a new process, rather than change in the technology which would displace [those people]'. From this perspective,
workers were not learning new skills, but were basically performing
previously acquired skills on new equipment.

Printers at the *Advertiser* believe, however, that demarcation and skills
issues were significant in the 1980s, and that substantial new work requirements were introduced. One view is that skill requirements have
decreased: 'Before it was more skillful. You could do in a day four or five
things you'd learned at trade school' (Patrickson 1986:5). Similarly, 'I
only use about 30 per cent of my old skills now – school leavers could
learn this in a few months' (1986:5). Commenting on the culmination
of technological and workplace developments in the 1980s and into the
1990s, a chapel representative suggested: 'The atmosphere of a trade has
all gone. You don't need to be a Rhodes Scholar to do cutting and pasting.
There's no atmosphere here any more. It's not a trade any more. It's just
a job.' Another chapel representative claimed: 'some of the artistry has
gone now. People are afraid to experiment, they don't have the artistic
skills. Once you've got an image on the computer bank, you can just pull
it up whenever you need it. No need for creativity. It's a pity.'

Some production workers argued, however, that the introduction of new technology has necessitated the learning of different skills, or the adaptation of old skills to new equipment and new conditions. One production worker compared the skills and conditions required at the old plant to those at Mile End. He said: 'I used to go home after a shift [at the old plant] and cough up about three million full stops. You'd look down the room and realise you were working in black mist. Mile End is a far cleaner place to work in. You still get dirty, but it just doesn't compare' (*Insider* 1992b:2). In terms of skill requirements, he suggested that the computer eliminates the guesswork associated with printing, and that 'with the modern technology down here, the machine does 90 per cent of the work. Unless something goes wrong, there's no running around like the old presses – everything's at your fingertips' (*Insider* 1992b:2). Nevertheless, duties, including operating the console for the print machine, handling paper reels and regulating colour all demand the possession of various new skills and training. This point was also made by a press maintenance manager who remarked that 'The work is far more involved these days than it was 20 years ago. Not so physically demanding but far more technical' (Deb 1994:1). Questions also arise over who is to work on new equipment, how such work is to be done, and how the equipment is to be operated. In the early period of operations at Mile End, considerable difficulties with the technology were experienced. Problems with the new presses included electrical and mechanical faults, motor difficulties, web breaks, and sensor detecting faults. Operating the new presses, according to the managing director at the plant, had put everyone on the 'fastest learning curves' of their lives (Scales 1991:1). Commenting on the process of adaptation and learning, a print worker claimed: 'The machines are like people in some ways. The theory is there, but you have to adapt them to your conditions. They all have differences. It's a matter of teaching them about using the machines.' He went on to suggest: 'It's a different type of skill that is needed here now. Some of the skills have been lost ... [But] it's still printing. The printers have a feel for it, can spot the problem. It may require a different solution, but it's still printers doing it.' The perceived gap between theory and operation was also noted by a press-room worker at Mile End who commented:

New technology has had an impact on us. We were operating the old equipment and we had to learn to use the new equipment and in between we had
to learn new skills. We had the manufacturers [of the technology] over here
but what they told us didn't work. We had to learn it for ourselves. The theory
didn't work. The operation was different.

While different skills were being learned, and many believed that learning and exercising these skills was less demanding, there was still pride in the work being accomplished. This is evident from the comment that 'our print wins best quality awards all the time; we win pan-Pacific awards. We probably do even better than the English [at Wapping], and they've had them [the presses] in for five years more than us'. For those working on the shopfloor, the introduction of new technology had an influence on their work, and on their relation to work and to other workers.

Inter-Union Relations

In addition to relations between management and the unions, the associations between the unions are an important element when considering issues of workplace reorganisation related to technological innovation. The new computer technology provides employers with more flexibility in hiring and work allocation practices. In particular, it provides a means whereby journalists and editorial staff can perform many of the jobs previously undertaken by print workers. Although it is primarily a management decision whether journalists perform such tasks, the possibility can strain the relations between journalists, and print and production workers, especially in cases where journalists and their unions support the introduction of technology which threatens the livelihood of the print and production workers. While relations between the PKIU and the AJA at the *Advertiser* have gone through various phases, they have been generally cooperative and supportive. A print worker, and union representative, commented:

> Our relations with journalists are good here, but they haven't always been that way. When I first came here, they wouldn't allow any overlap of roles; they had lost sight of the fact that we're all here to publish a paper. My earlier training and experience at a magazine was one where I had to do all parts of the trade. When I came here, I did some of the same things that journalists did. At first they didn't like it, but after I did it a few times and did it better, they got confidence in us, and we continued doing it. For example, chopping up stories to make them fit the space; sometimes you need to edit out parts and so on. Now we've got good relations with the journalists.

These observations were supported by a former journalist:

> I've always found [relations between journalists and printers] to be pretty good at the *Advertiser*. I suppose there was a period where the relationship wasn't as good as it used to be. It was pretty good in hot metal days, there were very different demarcation lines then; it was easy. I think there was a

slight change when the new technology came along. I think some composing room staff could see at that stage that some job functions might disappear, which had proved to be the case. But of late I think the realisation that the lines between composing and editorial have become so much more blurred, from my observations, the relationships have become much better.

When considering negotiation processes, relations between the various unions organised in a particular workplace are important. If unions can present a united front to management on issues relating to technological innovation and workplace reorganisation, management will find it difficult to proceed without taking into consideration the concerns of the workforce. When unions are divided, as was the case in England, management may find it easier to implement its programs by playing the unions off against each other. In addition, it would seem easier for unions to form a united front in a workplace involving only two or three unions, as in the case of the *Advertiser*. On the other hand, having only a few unions at the workplace can also benefit management in some instances, as a management representative explained:

> I think the *Advertiser* industrially has always been a bit different from other papers around the country . . . All our production employees were covered by the PKIU. We didn't have a proliferation of unions. We only had this one union arrangement with the production people. So that you had your journalists covered by the then AJA, all your production people covered by the PKIU, and the rest of the clerical staff were basically covered by a States Wages Bill decision which was effectively under PKIU representation again, but that was very much a flow-on of what happened in the production area. So I think that industrial organisation led to a far – I don't know if easier is the right word – but certainly a more comfortable relationship between the company and the employees and their union, because you were only dealing with one or two parties all the time. And I think historically out of that developed an approach of discussion and negotiation, and not head banging.

The existence of a small number of unions representing workers in the workplace, and with relatively good relations, has benefited both management and workers in the negotiation process. As MacIntosh suggested, 'In the long run . . . unionism has been strengthened, but . . . the unions have become less disruptive on small matters and easier to negotiate with as a group' (1984:36).

An analysis of negotiations at the *Advertiser* through the 1980s and into the 1990s concerning the introduction of new technology into the workplace shows that workers, management and employers had varying perspectives and different experiences of the technology. These perspectives and experiences influenced their understanding of the impact of the technology on the workplace. In particular, even if skill requirements were not raised by technology, they might be altered and subsequently affect job and labour requirements in fundamental ways. At the *Advertiser*, the introduction of technology into the workplace resulted in the disappearance of some tasks and work categories, and the creation of new skill requirements in other areas. The historically cooperative relations between management and unions meant that while there have been disagreements over these processes, and while the Commission intervened at various times, both parties influenced the relationship between technology and workplace reorganisation. Nevertheless, there is little doubt that the ability to influence negotiations in the workplace shifted towards management. The availability of labour-saving technology, the potential to employ an alternative workforce, and continuing high levels of unemployment provided management with a great deal of influence in negotiations at the workplace. The result has been that while the PKIU has been involved in negotiations, the number of its members employed has decreased through the 1980s and into the 1990s. The takeover of the *Advertiser* by News Limited in 1987 was to be another critical event in the shift in workplace power towards management, and was also to represent a shift in the culture of workplace practice.

Technological Innovation and Workplace Reorganisation: The Local Newspaper as Part of a Global Media Corporation

The *Advertiser* was purchased by News Limited in 1987 and negotiations about technological innovation and workplace reorganisation became not just local, but national and international. A senior management official spoke of the takeover as introducing:

> a feeling of great uncertainty to employees, and for the reason that they wondered about their jobs – where were they going to be, who the new manager was going to be, how they would tackle industrial relations ... whether they would adopt the same attitudes as *The Advertiser* would. It was a difficult 12–18 month period for us. Some decided they'd leave, take their retirement, weren't prepared to take any chances. Also, News Limited had different technology from ours in the sense that it was a different system, not different in the sense that it was basically different, but just it was a slightly later model of a different brand. Our people, *Advertiser* people, became uncertain as to whether they'd introduce this new system here and therefore what it would do to jobs here. As it turned out, eventually that later system was introduced here, but it was introduced after a period of time by which time everyone had settled down pretty well into accepting the fact that while it was different it wasn't a different sort of thing, in the sense that you had different owners who in some ways had different objectives, but not different in the overall sense, but different culture and style of doing things (see also PTJ 1987).

The sale of the *Advertiser* resulted in a new context for negotiation, the influence of which became evident when News Limited announced plans to move production of the *Advertiser* and various other newspapers to Mile End. While negotiations at the *Advertiser* previously occurred under a federal award, negotiation was essentially local in content, but now the newspaper was part of a much larger organisation. Negotiations at the *Advertiser* would now affect, and be affected by, negotiations and developments in other Australian cities and overseas. Both management and unions understood that the outcomes of negotiations at one workplace would set the standard for the other establishments. When News Corporation proposed to establish new production sites for their newspapers, negotiations in Adelaide became linked to those in other states, and in Britain. Discussions took place, for example, about plants using 'technology and methods similar to the Wapping site established in London, [and] the PKIU met with management to discuss many issues, including the proposed reduction by 50 per cent of employment of PKIU members' (PTJ 1990:13).

For both management and workers, new ownership exacerbated uncertainties relating to the restructuring of workplace relations. When negotiations occurred at the local level, participants believed that reorganisation could occur relatively smoothly, but this altered dramatically when other workplace sites became involved. One management representative claimed:

> Sometimes I'd say where you're introducing new technology there is always a pressure exerted on the local employees from some of the bigger interstate unionised employees. For example, sometimes when there's a dispute over new technology, over the introduction of new technology, where our people might be reasonably comfortable with it, if Sydney and Melbourne decide they've got a problem, it can spill over here without any real reason for it . . . [T]hat gets pretty frustrating sometimes.

While management and workers in Adelaide considered they 'knew each other' due to the relatively small size of the organisation, and of the city itself, the sense of familiarity was to some extent lost once 'outsiders' became involved. Discussing the shift of production facilities to Mile End, another management official stated:

> [The move to Mile End] was tougher than we had perhaps experienced in the past in the sense that we were then part of News Limited and we were the first plant of four to be built around the country, so whatever was negotiated for Adelaide was going to form the basis of what was negotiated in Melbourne, Sydney and Brisbane. Therefore, you weren't talking about negotiations that involved the *Advertiser* in its normal negotiation process with the union in terms of the *Advertiser* people. It was the *Advertiser* negotiating the

Advertiser conditions with everybody from Melbourne, Sydney, and Brisbane involved as well. A whole range of unions, a whole range of competing interests and claims ... So you really had a very, very detailed and elongated process, which was probably 12 to 15 or 18 months of actual negotiation. It was the sort of thing that we at the *Advertiser* believed – give us a week with our people full-time and we'd get it done. But we were dictated by the competing claims of all the various states.

He continued:

It was a detailed process. It involved a lot of very, very heavy meetings, but in the end result it was relatively amicable in the way it was done. You would have to say that there weren't stoppages, there weren't problems of that nature, there were times when there was ranting and raving, but no more than one might expect in the context of a very complex negotiation process.

For the owners and for management, the construction of the Mile End plant in Adelaide was considered to be a test for future developments in the eastern states of Australia. Management believed Adelaide was chosen for this purpose because of its smaller size, the fact that there were fewer people involved, and because of the historically cooperative relations between management and unions.

The influence of organisational size and knowledge of local conditions in shaping the content of negotiations were also noted by workers at the *Advertiser*. A journalist, commenting on the relative lack of industrial disputes in Adelaide compared with other Australian cities, suggested, 'It's a smaller city, smaller organisation, it's people knowing people, rather than being lost in a bigger organisation'. Comments from other workers suggested, however, that relations between management and the workforce had declined since 1987. One printer felt, 'It was a lot better before Murdoch bought us out in 1987. Now we're not working for the local *Advertiser*. We're working for Murdoch', while another commented that 'everything goes through Sydney [the location of the head offices of News Corporation Limited] since Murdoch came in; it makes negotiations difficult because [management] are always having to see what Sydney says'. As with management representatives, these workers believed that their ability to influence negotiations had been reduced by the takeover of the *Advertiser*.

The organisation of work at the *Advertiser* was also altered in a fundamental manner with the involvement of News Limited. Because News Limited owned newspapers in every major city in Australia, it had the opportunity to reduce costs by producing two or more publications at the same site. By reducing operating costs through sharing premises, computers, and other print and distribution facilities, the number of jobs available was reduced. This was made evident in a submission to the Australian House of Representatives Select Committee on the Print Media in which Professor Peter Swan, representing News Limited, argued that:

> Group ownership [by News Limited] also greatly benefits the national daily, *The Australian*, which utilises existing production facilities in Melbourne, Brisbane, Adelaide, Perth and Townsville after having been sent to these cities by facsimile from Sydney. Except for Melbourne and Perth, it is published along with the local product, saving cost in the overall shifts that would be required if both products were printed alone. *The Sunday Telegraph* is also printed by The Herald and Weekly Times in Melbourne in conjunction with their own *Sunday Herald-Sun* and Queensland Press in Brisbane prints *The Sunday Telegraph* along with *The Sunday Mail.*
> Moreover, in Townsville *The Australian* shares facsimile equipment (recorders) with *The Sunday Mail* sent from Brisbane and in Adelaide at the new Mile End Print Centre facsimile receivers or recorders are shared with all locally produced News Limited papers as well as the independent *The News*. Facsimile receiving equipment planned for the new print sites in Melbourne and Brisbane will also be shared with the *Australian* (AHRSC 1992:124–5).

The potential for cost-saving on the part of an organisation such as News Limited also extends to the distribution of newspapers, where for example:

> *The Australian* shares the use of delivery trucks with News Limited's local metropolitan dailies published in Melbourne, Brisbane and Adelaide. Adding *The Australian* to an existing delivery run is a negligible increase in cost due to the payment on the basis of mileage covered and hours worked rather than on weight as is the case with air freight (AHRSC 1992:125).

The savings for the company are again reflected in fewer jobs for distribution workers. Even journalists are under threat from technology when employed by an organisation with a reach as expansive as that of News Limited. Using stories in more than one newspaper, or sending company representatives to cover stories rather than journalists from each newspaper, has become a cost reducing technique on the part of newspaper proprietors (AHRSC 1992:124–31). In becoming part of a multi-site organisation such as News Limited which owns many newspapers, the impact of technology on workplace organisation at a single newspaper such as the *Advertiser* is heightened.

The context of negotiation at the *Advertiser* after the takeover by News Limited was also affected by events involving the international holdings of News Corporation. Negotiations over the move to Mile End occurred, for example, in the aftermath of the shift of News Corporation's newspapers in London from Fleet Street to Wapping. As indicated in earlier chapters, the reorganisation of the newspaper industry in London involved a long and intense struggle between the company and the print unions, resulting in News International using its financial power and the law to dismiss large numbers of workers and to establish a union-free workplace. Those involved in negotiations at the *Advertiser* were certainly aware of these developments, as is apparent from the following comments made by management and union members:

> There was always the comment that our plant at Mile End was Wapping South, . . . and other people said 'You know what those boxes have got out there. They've got razor wire from Wapping'. News Limited was very, very conscious of the, I guess, stigma attached to the Wapping exercise; very, very conscious that the Australian system was not that way inclined (management representative).

> The *Advertiser* was just progressing along, and when it all happened there weren't any papers burnt in the street [as in Wapping] or anything like that. It all happened with a lot of excitement, with a lot of hard work . . . We didn't get through it entirely without union pressures at certain times – we had our stop-works and so on – more a case of making sure the company did the right thing by the people rather than wanting to stop the technology coming in [as in Wapping] (management representative).

> People were saying we were going to have another Wapping here, but I didn't think that would be the case. We've never wanted to go the way of Wapping. We've always entered into negotiations (PKIU member).

> In Britain, they asked for it a lot themselves, SOGAT [Society of Graphical and Allied Trades] and the NGA [National Graphical Association]. We've always believed in negotiations. The technology is going to come. We've always tried to move in time with the technology (PKIU member).

Aware of events in England, the parties involved at the local level in Adelaide did not want similar events to occur there. The history of the relationship between the parties was an important element in ensuring that Adelaide did not become a 'Wapping South', at least in implementing workplace reorganisation. This relationship at the *Advertiser* was also affected by the manner in which both the PKIU and the AJA, with the support and direction of the Accord between the ACTU and the federal government, approached the issue of new technology. By adopting a broadly supportive attitude, while also demanding involvement in negotiations, the unions contributed to the relatively smooth process of implementation. Further, the presence of the Industrial Relations Commission was also important. The actors at the *Advertiser* had a history of negotiating under the guidance of awards of the Commission, intended to prevent disagreements between the parties developing into industrial disputes.

In June 1993 negotiations involving management from News Limited, its rival Fairfax, and unions in the newspaper industry, including those at the *Advertiser*, collapsed, however, resulting in a twenty-four hour national strike (PTJ 1993a). Early in 1993, newspaper unions across Australia, coordinated by the ACTU, were involved in discussions with both News Limited and Fairfax over wages and conditions, including demarcations over the introduction of electronic pagination into the production process. The wage increases and conditions offered by the companies were considered inadequate by the unions, resulting in stop-work meetings across the country in late May. Discussions followed in which the unions sought a national agreement and guaranteed minimum wage outcomes, while the companies wanted negotiations to occur with individual unions at the local level. News Limited, for example, issued a statement saying, 'We have told the ACTU and the unions that we are prepared to have discussions with our employees and their union representatives at each site. We are not excluding unions from the process. The days of pay rises for nothing bludgeoned by all-powerful unions have gone forever' (*Alliance* 1993a: 5). Unions rejected the approach of the companies, with the joint federal secretary of the Alliance stating: 'the company's position for site by site talks would mean that some people will be worse off' (*Alliance* 1993a:5). As discussions deteriorated, the first-ever national strike of all newspaper unions occurred on 4 June 1993, resulting in disruption of production and circulation. News Limited offered a 10 per cent wage increase over two years and agreed to lock itself into award restructuring. This initial offer carried with it the provision that unions accept all of the demands of News Limited concerning workplace restructuring. In practice this meant 'the unions had to agree to give the company open slather in removing demarcation lines and in the introduction of new technology, with no additional payments in the life of the agreement' (*Alliance* 1993d:3). The offer was rejected by the unions, and a new agreement was reached in which workers would receive a 10 per cent wage increase in five segments over two years. Agreement was also reached with the PKIU that training, retraining and redeployment would be available where possible, and was required in some instances, while it was also agreed that there would be no forced redundancies for two years. It was also agreed that demarcations rendered potentially obsolete by technological innovation would be eliminated over time (PTJ 1993a). Summarising the result of the dispute, the PKIU stated that it had 'reached agreement with News Limited over an orderly and negotiated process for the introduction of new technology, establishment of career paths and elimination of unnecessary demarcation' (PTJ 1993b:5). It was also decided that a national negotiating committee and a working party should be established to consider these and related issues. The Alliance obtained a similar agreement from News Limited, while also agreeing to use page make-up technology and a right to seek further money for extra skills required in that work, or in other work related to the introduction of new technology (*Alliance* 1993a, 1993b, 1993c, 1993d, 1994a, 1994b, 1994c; PTJ 1993a, 1993b).

This dispute was significant for a number of reasons. The combined action of the unions, under the leadership of the ACTU, put into practice the principles of strategic unionism. Different unions acted together in a context of restructured industrial relations practices, linked to issues including enterprise bargaining and the reorganisation of newspaper ownership (PTJ 1992). While News Limited wanted negotiations to be at a local level with individual unions, the combined unions considered this a threat to their attempts to establish minimum conditions across the industry. At the same time, the outcome of the dispute did not reverse the impact of technology on traditional demarcations, but at least it provided unions with some scope for influencing the eventual relationship between new technology and workplace reforms.

The dispute also illustrated further how management and union members can have different understandings of the same situation. A management representative at the *Advertiser* commented:

> Only recently the journalists had a 24-hour strike and about the last thing individual journalists here wanted to do was to go on strike; we know that, but it was a federal strike, not coming from here, and similarly with the printers, so it's not all plain sailing. I want to emphasise [that strikes or disputes might occur] not because you're not necessarily handling your local industrial relations adequately, but because there might be a real perceived problem interstate [and] because of federal awards, the pressures spill over into the other smaller states.

Subsequent events cast doubt on this interpretation of events, and suggest that workers at the *Advertiser* were capable of taking action on their own initiative without interstate pressure. On 15 November 1994, members of the Alliance at the *Advertiser*, with the support of PKIU members, attended an unauthorised stop-work meeting to discuss the offer by News Limited of non-union contracts to Alliance members, and were subsequently locked out. On seeking re-entry into the *Advertiser* building, members involved in the meeting were refused entry unless they signed a statement to the effect that they had attended an unauthorised stop-work meeting, and further, that they would agree not to partake in any further industrial action. Each member was told they would be dismissed until further notice unless they signed the statement. This then led to displays of solidarity, including PKIU members locking themselves in the fax room to prevent the faxing of pages to the Mile End production plant (*Alliance* 1994c). Such instances of industrial action suggest that relations at the workplace have been affected by the new ownership of the *Advertiser*. The right of unions to represent the workforce was no longer to be taken for granted, and had to be defended by the unions.

Technological Innovation and Workplace Reorganisation at the *Advertiser*

Technological innovation and workplace reorganisation occurred through the 1980s at the *Advertiser* primarily by negotiation. As the evidence presented here illustrates, however, the negotiation process was not one into which all parties had equal input or control over decisions. The different positions of the respective parties in the organisation meant that the 'same' processes had different consequences. A retraining program, for example, that was successful from the perspective of management might constitute a threat to the livelihood of those who needed retraining. Similarly, the meaning and impact of redundancy might appear self-evident, but can be interpreted differently by the various actors. In announcing its plan for new production sites across Australia, News Limited recognised that there would be up to a 50 per cent reduction in its workforce, with PKIU members especially being affected. According to News Limited spokesperson, Tom Stokie: '[T]he redundancies would not be compulsory but voluntary', and he added that 'staff are currently being trained at overseas sites on the new presses and platemaking equipment' (PTJ 1990:13). For management, the redundancies that eventually occurred were part of a well-executed plan to modernise production. According to a management representative at the *Advertiser*, redundancy programs for print workers were part of:

> a process of change which has generally been well planned, well documented, well argued if you like, and well sold, and . . . we have had the benefit in recent times of being able to reduce the numbers where necessary on a voluntary basis, so certainly we are employing less, but a lot of the lesser number have come from people who have taken the opportunity two or three years earlier than they might readily have done, because the bonus has been that good, and as such, that has made it a more comfortable arrangement as well. We haven't had the necessity at this stage to go around and finger people and that makes a lot of difference in the overall scheme of things as well.

From this perspective, those who had taken redundancy did so of their own choice, and in response to the generosity of the organisation.

For the individual worker, by contrast, redundancy was associated with uncertainty about their future in the newspaper trade. According to a chapel representative:

> I can't see the newspaper industry lasting much longer. They always said 2000 was the year for newspapers to end. We'll see ... People [i.e. union members] ask me if they can afford a car. They ask if they'll have a job in two years. I can't tell them. I just tell them to enjoy it while they've got it. I can't tell them what's going to happen.

Uncertainty concerning the future may contribute to workers accepting redundancy payments which they might not take in other circumstances. As one worker claimed, 'If you offered people redundancies, many of them would take it straight away. We're trying to push for redundancies based on six weeks [pay] and management tells us they've got thirty guys knocking on their door accepting four weeks.' There was a diminished belief in the ability of the union to pursue claims, linked to the increased ability of management to replace and train workers. The different location of workers and management in the labour process influenced their capacities to make decisions. For management, redundancy payments were a way of managing the attrition that comes with technological investment; for the worker, accepting redundancy can be a response to uncertainty – that is, the concept of 'voluntary' is constrained by structural conditions and workplace reorganisation.

The practice of negotiation itself has built into it power relationships that come both from the shopfloor and from societal relations. Even though both sides agree that peaceful negotiation continues to predominate at the *Advertiser*, the content and results of such negotiation carry different meanings and consequences for the parties. Workers can agree, for example, that '[o]ur unions have done better in negotiations than has Sydney or Melbourne ... We've had better people here than over there – willing to try to work things out'. Similarly, in relation to electronic pagination, one compositor said, 'We're going to negotiate about it. Try to get protection for our members and ourselves. Generally we've been OK – so far!' Another compositor agreed, saying '[W]e'll move to pagination eventually, can't stop it, it's inevitable, but we're trying to negotiate, to protect our members'. This perception was to be crucial in the reorganisation of the workplace that accompanied the introduction of electronic pagination. On the other hand, colleagues claimed, 'We've had to fight for everything. Management won't give anything without making it a fight', and 'The idea of consultation is all bullshit. It's the same as it's always been. They just tell us what they've decided and we have to fight for whatever we want.'

While there was willingness to participate in negotiations, there was also recognition that the participation of unions in negotiations was limited by factors including the control that proprietors and managers exercise over investment decisions. That is, the relationship between technological innovation and workplace reorganisation was mediated by the relations that developed between management and workers, and influenced by the broader national and global contexts.

Theoretical Lessons of the *Advertiser*

The case of the *Advertiser* provides an example of workplace reorganisation being shaped by local, national and international conditions. During the 1980s at the *Advertiser*, there were significant developments in technological hardware and a shift in the location of the production plant, accompanied by extended periods of negotiation between the relevant actors. While the owners, management and workforce all contributed to these negotiations, management had a dominant role in most situations. One factor important in explaining why all parties were able to contribute to negotiations – especially up to 1987 – was the relatively small size of the firm. The issue of firm size cannot be understood, however, in isolation from the institutional and societal context. At each stage in the developing relationship between the introduction of new technology and workplace reorganisation, the Industrial Relations Commission had either been called on to make rulings over issues, including the influence of technology on workplace skill requirements, or was an institution to which the parties could be compelled to turn in the case of dispute. The very existence of the Commission influenced the way that workplace issues were negotiated. The role of the Labor government and of the ACTU, in conjunction with the PKIU and the AJA, was also crucial to the negotiation process. By advocating strategic unionism, and attempting to create a model of industrial practice based on consultation and exchange of information, the federal political and union organisations provided support for the approaches of both the PKIU and the AJA. This was in marked contrast to the earlier experience at Fairfax newspapers in the 1970s where no such policy was pursued. The unions involved in the dispute with the Fairfax management were isolated and unable to draw on the support that existed through the 1980s. The ability of the unions to organise workers in the workplace at the *Advertiser*, and the cooperative relations between the AJA and the PKIU, also gave labour a united presence with which management and employers at the *Advertiser* had to negotiate. Modification of the industrial relations system, technological developments, and workplace reorganisation were also affected by policies of economic restructuring, which were themselves shaped by global economic factors. These factors included the global dispersal of information technology, high levels of unemployment, and policies of free trade and economic liberalisation.

Interactions between local practice and the globalisation of information and media technology and corporations were also apparent in the process of reorganisation of the workplace at the *Advertiser*. The company was able to base its plans for improvements in its Australian newspaper holdings on its experiences in England, particularly in terms of the type of technology it sought to introduce. Both the company and the workforce were able to learn from the experiences of their British counterparts in operating new production technology, although this did not mean that there were no problems in introducing the technology in Australia. In terms of negotiations with the workforce, the influence of local practice was very evident. While new owners and new management became involved in workplace negotiations from the late 1980s onwards, the historically developed local structures for negotiation meant that neither party sought to repeat the events at Wapping. Nevertheless, the strike that occurred in the early 1990s involving members of the PKIU at the *Advertiser* suggested that some of the overtly confrontational managerial practices that existed in London might be appearing in Adelaide. With the election of the Liberal–National Party coalition to federal government in 1996, and its subsequent legislative activity intended to reduce the influence both of the Industrial Relations Commission and of trade unions, it remains to be seen whether conflicts such as those at Wapping will develop in Adelaide as there occur further technological and workplace developments. Indeed, News Limited has continued in its attempts to introduce individual contracts of employment, which to date have been resisted by the unions.

Conclusion

This analysis supports the relational model which recognises that structural constraints are important in providing a context in which state institutions and actors influence the relationships among new technology, social relations in organisations, and workplace reforms. Even though practices of cooperation, and a strong union tradition, existed at the *Advertiser*, management retained the initiative in introducing workplace reorganisation. This outcome was related to both the financial power of management, particularly with the resources invested by News Limited, and to the labour-saving potential of the new technology, which placed labour in a defensive position for much of the period under analysis. The role of News Corporation also meant that events in the global economy had a direct influence at the *Advertiser*. The international pursuit of technological advances by the company meant that developments at the *Advertiser* took into account earlier events at Wapping, while they were also being observed for possibilities for future developments elsewhere in Australia, Britain and other parts of the company. In addition, while the Accord and associated policies of the government were introduced in part to empower unions in the workplace, they were also introduced as a means of opening up the Australian economy to world competition. As the government attempted to attract international investment, it sought restraint from unions by promoting consultation and negotiation as opposed to dispute and confrontation, which constrained the modes of activity available to unions as they developed a partnership with the federal government.

The case study of the *Advertiser* supports the proposition that relationships between the introduction of new technology and eventual workplace reorganisation are dependent on the prevailing balance of power among unions, management and employers. It also demonstrates that those relationships are influenced by industrial contexts shaped by state institutions and actors such as industrial commissions and governments, and by national and international economic and social relations. In particular, the *Advertiser* case study indicates the need to examine international influences on local workplace reform and suggests that the relational model needs to be modified to take into account the effects of globalisation on technology and industry-wide workplace reorganisation. International contexts are explored further in the next chapter, where an analysis of the activities of News Corporation in the United States is presented.

CHAPTER 7

News Corporation in the United States: The Land of Opportunity?

Statements made by Rupert Murdoch in lectures to US audiences over a number of years have made clear the attraction of the United States to him both as an individual and as a site for extensive investment by News Corporation. Murdoch's perception of the United States as a 'land of opportunity' for businesses such as News Corporation with financial capital to invest, combined with the position of the country at the forefront of technological innovation, with a labour market not constrained (from the perspective of the company) by excessive union and worker intervention, all come together to make the US a very attractive proposition (Murdoch 1986b, 1989a). True to Murdoch's perceptions of the United States, the company has become heavily involved there, with extensive investments in all areas of the media, and on a more personal note, with Murdoch becoming a US citizen.

In the 1980s, the News America Publishing Inc. (News America) subsidiary of News Corporation Limited became increasingly visible and influential in the US media industry. According to one prestigious media publication, in the 1980s and 1990s 'Rupert Murdoch was the most conspicuous of the high rollers in the media acquisition business' (Emery, Emery with Roberts 1996:594). The company bought the 20th Century Fox studios for US$575 million, seven major television stations for US$1.55 billion, Harper and Row book publishers for US$300 million, and Triangle Publications, including *TV Guide*, for US$3 billion. In addition, the company became heavily involved in cable television and US sport. The newspaper interests of the company, as documented in Figure 7.1, have varied since News America's purchase of three newspapers in San Antonio in 1973, which subsequently became the *San Antonio Express–News*. At various times, News America has owned the *New York Post*, the *Chicago Sun-Times*, and the *Boston*

164

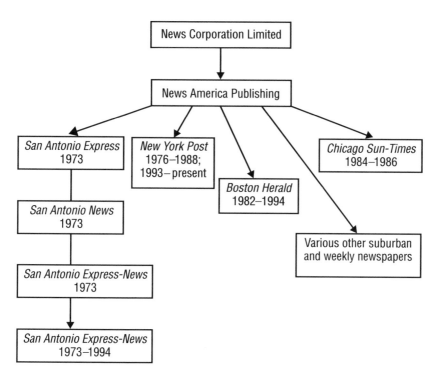

Figure 7.1 Major Newspaper Interests of News America Publishing, Inc.

Herald, while also being involved in bids for and negotiations over newspapers, including the *Buffalo Courier* and the New York *Daily News*. While being noted in the US newspaper industry for the so-called sensationalistic product of the company, News America has also been involved in negotiations and disputes over workplace reorganisation, related in some instances to technology. Given that many newspapers in the United States had invested in technology prior to the entry of News Corporation into that country, the company was not as heavily involved in technology breakthroughs in the newspaper industry in the United States as in Britain and Australia. Nevertheless, negotiations over technology did occur, while bargaining over other workplace matters was influenced by issues related to new technology.

The *New York Post*

The newspaper with which News America and Rupert Murdoch are most closely associated in the US is the *New York Post* (the *Post*). While

News America purchased its San Antonio newspapers before it bought the *Post*, the national exposure the company attained by owning a newspaper in the most international city in the US brought the company national and further international recognition. The startling headlines and the editorial direction of the newspaper reflected the established practice of the company, while the practice of the company in its negotiations and dealings with the New York newspaper unions was the subject of much discussion even among the battle-hardened members of the US newspaper industry. According to Gene Ruffini of the New York Newspaper Guild, for example, the twelve years in which News America owned the *Post* from 1976 to 1987 were the 'most turbulent and bitter period of union–management relations' in the history of the company (Ruffini 1989:7). For News America, which repurchased the newspaper in 1994, the *Post* has never been a profitable paper. In the mid-1990s, Murdoch claimed that News America had lost more than a hundred million dollars in its years of owning the *Post* (Auletta 1997). What the *Post* does give News America is a powerful voice and presence in a city which is central to the new information economy.

By 1976, the *Post* had been published without a break for 175 years, and since 1939 had operated under the ownership of Dorothy Schiff. In the early 1960s, a major strike of newspapers in New York ended for the *Post* when Schiff reached a settlement with the unions independently of the other New York newspaper proprietors. As a result, the *Post* emerged as the only afternoon daily newspaper in New York, and with an increased circulation (Emery, Emery with Roberts 1996). As part of the strike settlement with the union, the *Post* had tried to introduce computers into the typesetting process, but was unable to secure union agreement. In the early 1970s, the circulation of the *Post* began to decline in the face of competition from newspapers such as the New York *Daily News* and the New York *Times*, which also secured a larger share of advertising, and the crucial finances that come with such support. By 1976, the *Post* was losing around US$50 million per year (Shawcross 1992a). Schiff realised that to make the newspaper profitable she would need to invest in new plant and machinery and reduce labour and distribution costs, while also securing more advertising (Tucille 1989). In this context, she became willing to sell the newspaper, and entered into negotiations with Murdoch who was looking to secure a New York newspaper. In November 1976 Murdoch bought the *Post* for US$32.5 million (Tucille 1989). Soon after purchasing the newspaper, Murdoch sought to assure the unions involved at the *Post* that News America was not anti-union, and intended to honour the existing contract (Ruffini 1989). According to Ruffini, who was a journalist at the *Post*, Murdoch also claimed that as long as the staff were able to make the paper viable and successful, they would have nothing to fear (Ruffini 1989). Relations between the company and the unions, however, were soon fraught with hostility. Almost immediately on purchasing the newspaper, News America began to import labour from elsewhere in the US, as well as from Australia and Britain, to fill new journalistic positions with job titles such as 'special contributor', 'special contractor', and 'special correspondent'. Importantly, such positions were created with the claim that they were executive positions, and beyond the scope of coverage by the Newspaper Guild. As such, these appointments could be seen as an attempt to build a non-union journalist workforce. With these moves, the attempts of News America to circumvent the historically powerful New York newspaper unions had begun.

In 1978 News America, along with management at the *New York Times* and at the New York *Daily News*, were involved in negotiations with the New York newspaper unions over a new contract. These negotiations, which were linked to issues including technology and cost cutting, were ultimately to result in a major strike, involving up to five unions at one or more of these papers. The unions who took strike action were the International Printing and Graphic Communications Union (the pressmen's union) and the Newspaper Guild, as well as the Paper Handlers' Union and two locals (or branches) of the International Association of Machinists. The print unions came together in the strike as the Allied Printing Trades Council. In addition, the International Mailers' Union threatened to take strike action if the publishers resumed publication without reaching a settlement with the other unions (*Guild Reporter* 1978b). The basis of the 1978 strike could be traced to 1974, when the then ownership and management of the *Post* had been successful in introducing photocomposition technology into the workplace with the agreement of the print union. As part of the negotiations, linotype operators gave to management the right to decide how many operators were required, in return for an agreement not to lay off anyone currently working as a printer. This agreement, however, covered only the linotype operators and not the operators of the printing presses, who were represented by another union. Similar developments occurred in 1974 at the *New York Times* and the New York *Daily News*. As labour reporter A.H. Raskin wrote, the agreements reached in 1974 meant that 'the way was cleared for operating changes that spilled out of the composing room into all other parts of the papers' (1979a:41–2). The potential consequence was that the New York City newspapers could soon be produced without print workers and without unions, and it was the 'spilling out' identified by Raskin and linked to technological innovation that threatened to occur in 1978.

To implement further their program of modernisation of production and reduced labour costs, in 1978 News America and other New York City proprietors directed their attentions to the press room. Management at the *Post* and at the other newspapers wanted the right to determine the numbers operating the printing presses in the press room, in return for providing a life-time job guarantee to those already employed. Indeed, the plan was to reduce the overall staff levels by 50 per cent of the 1550 press operators at the *Post* (*Guild Reporter* 1978b). Arguing that the press rooms were overcrowded with unnecessary workers, and that press operators were working for only 15 per cent to 30 per cent of the time on each shift, the publishers proposed to alter the means of staffing these rooms from unit manning to room manning. Under unit manning, a newspaper was bound by contract to assign a fixed number of operators to each press and its associated functions. Under room manning, the company would be able to deploy its workforce as it saw fit, while promising life-time employment security to current workers. Management also proposed that the number of press operators (journeymen) employed on each press be cut from twelve to eight, again with various shift and work guarantees for those who remained. In terms of potential savings and costs from adjusting to this new system, News America had the comparative example of their plants in San Antonio. According to Raskin:

> At the *Post*, the basic crew on each press consisted of seventeen journeymen and juniors; in San Antonio, the number had been fixed by management at seven. The cost in direct wages and fringe benefits of maintaining each New York pressman averaged roughly thirty thousand dollars a year. On that basis, the institution of room manning here could mean a reduction in the annual labour bill for each standard press from five hundred and ten thousand dollars to two hundred and ten thousand. And comparison with the Texas papers indicated that similar savings were possible in the crews assigned to maintenance and cleanup. If the total reduction came to fifty percent – a goal that Murdoch considered quite feasible – the three New York publishers could look forward to a collective saving of well over twenty million dollars a year in their pressrooms (1979a:66).

Negotiations stalled as press operators realised that this plan would lead to the elimination of their jobs at a tremendous rate. The press operators proposed that only one press operator should be cut from each press rather than the four demanded by management. During these negotiations, Rupert Murdoch was elected president of the Publishers' Association of New York City, which was established in an attempt to provide a unified front for the proprietors. The determination of Murdoch to reshape working conditions at the *Post* was driven partly by the view attributed to him that 'over the past forty years *Post* publishers had been demonstrating their "liberalism" by abandoning management prerogatives to the unions and saddling the paper with high-priced deadwood' (Raskin 1979a:50). He was also determined to make the most of new technology, and even to seek the possibility of establishing a non-union workplace as in San Antonio. Murdoch's plans for reorganisation were met with opposition from the unions and as it became evident that no terms of agreement could be reached, on 9 August 1978 the press operators went on strike with the support of the Guild and other craft unions (Raskin 1979a, 1979b). As the strike progressed the parties sought a settlement, although they were unable to agree on terms. At one point, the company reviewed its initial plans and made a proposal which, according to the union, meant that at the *Post* 522 workers would be dismissed, of whom only 100 would be through attrition (*Guild Reporter* 1978b). According to the company, however, the new proposal meant that all but 200 of the 1550 'regular' press operators would be guaranteed jobs. As these talks floundered, the unions sought the assistance of Theodore Kheel, a long-standing mediator in newspaper labour disputes. An expanded role for Kheel as a mediator was rejected by the proprietors who saw him as having lined himself up with the union. The dispute continued, with rumours emerging that the publishers of the New York newspapers were possibly going to publish a single newspaper with non-union labour, while evidence was provided that earlier in the year the *Post* and the *Daily News* had both published 'practice' papers using non-union labour, and that Murdoch had brought in non-union labour from San Antonio (*Guild Reporter* 1978c).

As the dispute progressed, various alliances and agreements became strained. News America was uncertain as to whether the owners of the *Daily News* and the *Times* would alter their plans for employment reductions. The result was that on 28 September, News America broke from the proprietors' alliance and reached an agreement with its workforce to provide them with the terms ultimately agreed to at the other newspapers concerning employment levels. At the *Post*, the staffing agreement reached meant that the 1550 press operators would have their jobs guaranteed, with job reductions to occur over six years through attrition. As the strike continued at the other papers, the *Post* took advantage of its position as sole publisher in New York City, publishing a million copies of an advertisement-filled 128-page edition. In contrast, its pre-strike newspaper had been around half the size with about 625,000 copies published. Meanwhile, negotiations between the unions and the *Times* and the *Daily News* continued, although very little progress was made. The pressmen's union was fearful that they faced derecognition, as had occurred at the *Washington Post*, while the proprietors were concerned that they faced financial ruin if they did not significantly reorganise the workplace and reduce staffing levels. Eventually, the pressures of intensive negotiation and the financial needs of the union members began to tell, and one by one the unions returned to work, with various job guarantees in exchange for control over staffing levels and work practices. The *Times* and the *Daily News* recommenced publishing on 6 November. While the proprietors were able to achieve immediate cuts in maintenance and clean-up operations, in the final outcome 'the eighty-eight day strike had not produced the savings the publishers originally sought – only one fewer pressman was to be assigned to each machine, though there were more significant reductions in apprentices, maintenance men and supporting personnel' (Leapman 1983:120). According to William Kennedy, then president of the New York Pressmen's Union, the settlement was 'just and equitable', and meant that 'we didn't have a *Washington Post* in New York' (*Guild Reporter* 1978d:7).

At the same time that management at the New York newspapers had been negotiating contracts with the print unions in 1978, they had also been negotiating with the journalists in the Guild. As these negotiations with the Guild developed, the new management at the *Post* were soon making demands that the other proprietors were not prepared to follow. In February 1978, News America presented a list of 148 Guild members it wanted to dismiss which amounted to one-third of Guild members at the *Post*. Then in March 1978, the company devised what became known as the 'Auschwitz clause' (Leapman 1983:111). This referred to an announcement made by the company which stated that 'The *Post* proposes, on a one-time basis, the right to terminate members of the staff, with appropriate severance pay, who in its judgment are incompatible with the new management's publishing concept' (Leapman 1983:111). To this point, Murdoch had been negotiating with the Guild in concert with the publishers of the *Times* and the *Daily News*. These latter demands were considered, however, to be going too far. While the other proprietors were also concerned with cost issues, they believed that the conflicts such demands would raise with the Guild, and possibly with other unions, would outweigh any benefits that were gained by reducing the staff (Raskin 1979a).

A union perspective on both the so-called Auschwitz clause and relations in the newspaper industry more generally was put forward by Charles A. Perlik, the president of the Guild, who claimed:

> [T]he real pace setting [in contract negotiations] seems to be going on, as always, in New York, where the *Daily News*, the nation's largest-circulation newspaper; the *Times*, its most prestigious, and the *Post*, perhaps its splashiest,

have banded together in what is, for New York at least, clearly a well-coordinated plan to make the term 'contract improvements' acquire an entirely new meaning.

The *Post*, under the excuse of financial extremity, proposed what has been dubbed 'the Auschwitz Clause' – the untrammelled right to fire up to 35 per cent of its employees whom publisher Rupert Murdoch considers 'incompatible' with his publishing philosophy.

Aside from this flare of originality, Murdoch teamed up with his publishing colleagues to demand dozens of retrogressions in key Guild clauses on sick leave, job security, automation, subcontracting, health and welfare, overtime, scheduling, trial periods, the work week, retirement, rehiring and grievance procedures, among many others. Among the demands they put forward in a time of snowballing inflation was a freeze on wage minimums in lower classifications (Perlik, quoted in Bingel 1978:14).

As the threat of a strike by Guild members increased in response to these demands, the company began to fly in an alternative workforce from San Antonio and elsewhere to show that the writing and production of the newspaper would continue in the event of any stop-work action by the Guild, even if supported by production workers. Despite such threats from management, when the press operators went out on strike on 9 August, the Guild observed their picket lines. On 22 August, the Guild itself went out on strike when the *Post* took out a court suit trying to take away the right of the Guild to arbitrate grievances (*Guild Reporter* 1978a). After intensive intervention from the Federal Mediation and Conciliation Service in Washington DC for two months, terms were reached in which 122 out of 460 Guild members at the *Post* accepted severance payments, while eighteen less senior Guild members were laid off. The Guild managed to secure job security protection for the remaining staff, in addition to arrangements for a flexible working week and some wage increases (*Guild Reporter* 1978b; Shawcross 1992a).

For management, the source of the 1978 strike involving the newspaper unions was the intransigence of press operators and other production workers protecting expensive and technologically inefficient workplace production and organisation practices. At least at the *Post*, such workplace practices were claimed by management to be having an adverse affect on profits, with the paper running at a loss. From this position, the need for drastic action was considered crucial. In response to the charge that he had acted ruthlessly in the dispute, Murdoch stated that while the actions taken could be seen as ruthless, the company had been forced to act at the *Post* to save the business. He claimed further that a combination of abuses of union power and soft managerial practice in the past had made action in the present necessary, and that such action included cuts in the workforce (Murdoch 1988).

In its analysis of the origins of the dispute, however, the unions argued that what had really occurred was that management 'presented demands and took actions which they had to know could have no other result than that which occurred: the pressmen walked out' (*Guild Reporter* 1978a:7). Central to these demands was the introduction of technology, accompanied by substantial reductions in the workforce. As reporter A.H. Raskin wrote in his analysis of the dispute:

> What clearly emboldened the New York publishers now [at the time of the strike] was the extra muscle provided by a technological revolution that makes it possible for a comparative handful of executives, confidential secretaries and other non-union personnel to prepare and publish newspapers even with all the printing crafts on the picket line (1978:198).

That these concerns were important to newspaper workers involved in the dispute was made apparent in an editorial in the ITU journal, in which ITU President Bingel in conjunction with other senior officials of the ITU had the following to say about the dispute:

> Currently we are enjoined in a battle on a number of fronts in both the United States and Canada. The most expensive fights are the result of employer attempts to obliterate our union or another union in our industry.
>
> Not content with near-monopoly control of the newspaper industry in the United States and Canada, nor with swollen profits made possible by automation and its resultant reduction in costs, publishers across the country have launched a tough campaign against their employees. Adopting a technique of reverse collective bargaining, they have been successful in not only reducing the *number* of employees, but actually rolling back long-established conditions which have been achieved over many years of negotiations. Emboldened by these successes and encouraged by the apparent lack of concern for such tactics by the general public, they are attempting to undercut all the unions with whom they negotiate by wiping them out financially.
>
> New York has become the classic example of a publisher plot to destroy unions. Following the *Washington Post* example, the pressmen were selected as the target. In so-called negotiations, proposals were made to the pressmen which, if accepted, would virtually wipe out that union. When the pressmen declined to commit suicide, conditions of employment were posted by the publishers which, if agreed to, would have made the union members company property. The pressmen struck . . . The printers, having made many concessions, were out. All other unions were told to get lost until the pressmen could be brought to heel.
>
> Next came the pressure phase. According to the employer plan, with all unions pouring out benefits, the squeeze on the pressmen was to prove irresistible, thus shattering the union front . . . a classic divide-conquer technique calculated by the management generals to take about three weeks (Bingel et al. 1978:4–5).

Also making the link between the disputes at the *Washington Post* and the *New York Post* was the president of the Newspaper Guild, who commented in an address to the forty-fifth annual convention of the Newspaper Guild on 26 June 1978 that:

> [The publishers] are becoming even more aggressive. If you're looking for the place where it started, you need search no farther than the *Washington Post*, which announced in the course of bargaining for a new contract that it was no longer going to observe some of the key provisions of the expired one. New York's Big Three took it a step farther by cancelling our contracts wholesale following expiration and announcing that they would no longer be bound by the old terms and conditions during bargaining for new ones (Perlik, quoted in Bingel 1978:4).

From this perspective, the strategy of management in the dispute was to remove the press operators as a unionised presence from the workforce. Such a strategy entailed a number of tactics, including the introduction of technology, a united front on the part of the proprietors, and attempts to divide the unions. In addition, at least from the perspective of the union leadership, the example of the *Washington Post* was still vivid, and was being used as a model of how to proceed with technological innovation and workplace reorganisation by the New York proprietors. While a settlement was eventually achieved, it eventuated only after dramatic and costly confrontations between management and the unions. The historically developed organisational strength of the newspaper unions in New York meant they were able to resist, at least partially, the attempts by management to reduce their influence in the workplace. Nevertheless, management had succeeded in gaining some concessions from the unions, and this was a trend that was to be followed in the 1980s and 1990s.

In the aftermath of the dispute at the *New York Post*, and after a number of other conflicts through the 1980s, Murdoch still felt that the unions exercised too much influence in the workplace. In a speech delivered at the Manhattan Institute, he suggested in particular that unions were costing newspaper publishers money by forcing them to retain jobs that could easily be mechanised. He went on to say that there was a temptation for well-established publishers to give into union demands and attempt to absorb such unnecessary costs, reasoning that smaller competitors would not be able to do so. Murdoch stated, however, that this practice was a source of grave problems, and would ultimately lead to a confrontation between publishers and unions (Murdoch 1989a). For Murdoch, even in the late 1980s the unions in the New York City newspaper industry were still too influential in the workplace, and still had too much capacity to disrupt production, thereby preventing proprietors from producing efficiently and profitably. The repercussions of this perception were to re-emerge in the 1990s.

In 1988 Murdoch had sold the *New York Post* to Peter Kalikow because of legal restrictions on multi-media ownership, which prevented ownership of a newspaper and television station in the same city. There followed an incredible sequence of events relating to the ownership of the *Post* in which Kalikow, financier Stephen Hoffenberg, and parking garage magnate Abe Hirschfeld all assumed control of the newspaper at various times. While the *Post* had not been profitable under the ownership of Murdoch, the resources of News America meant the paper could continue publication. Under the succession of owners that followed Murdoch, however, and despite winning various concessions from the unions, the survival of the *Post* was no longer guaranteed. In 1992, for example, the *Post* lost more than US$7 million, and into 1993 was losing US$200,000 to US$300,000 a week. In addition, circulation had plummeted from more than 900,000 under Murdoch's ownership to around 400,000 (Garneau 1993). With the newspaper declared bankrupt in early 1993, Murdoch re-emerged as the saviour of the *Post*, being granted a Federal Bankruptcy Court order allowing him to take over management control for sixty days while also reimbursing the existing publisher. During those sixty days he had to seek a waiver from the Federal Communications Commission on the cross-ownership rule that had originally required him to sell the paper. Murdoch was granted a permanent waiver in June 1993 by the Federal Communications Commission (Emery, Emery with Roberts 1996). On being asked why News America would want to buy a loss-making newspaper, Murdoch told *Post* staff that publishing was about improving society, not about making money (Garneau 1993). The executive vice president of News Corporation was to add that this was a real opportunity to own a newspaper in 'the world media capital' of New York, and that the financial issues were insignificant (Garneau 1993:10). As before, however, News America made it clear to all concerned that they would not buy the paper unless 'appropriate agreements' regarding the financial feasibility of the newspaper were reached with the unions and creditors (Garneau 1993:10). News America and Murdoch threatened several times to abandon plans to repurchase the *Post* unless the unions agreed to major concessions. Indeed, the paper was closed down from 9–11 July and was only re-opened after the unions gave up US$6.2 million in concessions (*Facts on File* 1993a, 1993b, 1993c; Garneau 1993). The paper was again shut down on 27 September when the Guild went on strike, with the support of the other unions. In this instance, the Guild went on strike in response to Murdoch's demand that the *Post* have the right to fire Guild members at will, while the *Post* also refused to pay those affected more than a few weeks' severance payment. Three days later, however, the production unions returned to work, and the Guild then voted to return to work. The Guild members were required to re-apply for their jobs, and only thirty-five of the 270 who had gone on strike were rehired (Kurtz 1994:348). Even after the conflicts of the 1970s and 1980s, News America was not content to leave workplace conditions as it found them. In contrast to disputes in the 1970s and 1980s, in this instance the Guild was crushed. In the words of reporter and commentator, Howard Kurtz:

> In an era when corporate owners routinely walk away from money-losing properties, the *New York Post* limped sadly on, a symbol of one man's willingness to subsidise a failing newspaper once he had obliterated its staff. Whether such a fate is preferable to death with dignity is not an academic question for those of us who have been through such a wrenching experience (1994:348).

The question of death with dignity was one that confronted workers at the *Buffalo Courier* which News America attempted to purchase in the 1980s.

The *Buffalo Courier* and the *Boston Herald*

In the early 1980s News America was involved in attempts to purchase two financially struggling newspapers, one in Buffalo, which was unsuccessful, and the other in Boston, which was successful. While News America was ultimately unsuccessful in purchasing the *Buffalo Courier*, the negotiations surrounding its attempt to buy the newspaper from Cowles Media in 1982 are illustrative of the engagement of the company with the US newspaper industry. In 1982, Cowles Media announced that the size of its reportedly significant pre-tax losses of the previous three years meant that publication of the *Buffalo Courier* would cease after the edition of 19 September 1982 unless a new buyer could be found. Among the 1100 employees of the *Courier* who were threatened with job loss if this closure occurred were 160 members of the ITU and 370 members of the Newspaper Guild. In total, nine unions represented 670 employees at the newspaper (*Guild Reporter* 1982a, 1982b, 1982c).

News America announced that it would be prepared to buy the newspaper provided agreement could be reached with the relevant unions over workplace issues. News America would pay nothing for the assets of the newspaper, but would assume 'most' of the liabilities. News

America demanded both a 40 per cent reduction in staffing levels, and the freedom to choose which employees were to be dismissed. On 13 September 1982, News America presented the unions with a fourteen-point proposal, and imposed a deadline for a final decision of midnight on 16 September. The fourteen-point proposal from News America contained the following demands:

1 Reduction in the workforce by at least 40 per cent. This would involve cutting the 156-person editorial staff back to not less than ninety employees, and the company further would only guarantee that seventy current editorial department employees would be retained, and envisaged further reductions through resignations from among those seventy employees. In addition, it was proposed that the number of mailers' jobs be reduced from fifty to twenty-four; printers and stereotypers to be reduced from 147 to fifty-five; inside classified sales persons represented by the ITU to be cut from twenty-two to nine; press operators to be reduced from sixty-six to forty-one; and machinists from nine to five.

2 An extension of current contracts by one year in order to allow a 'stretch-out' of wage increases. There would be no wage increases in the second year of the contract. The wage increase that had been negotiated for the second year would be moved to the third year, and the third year increase to the fourth year.

3 Elimination of all staffing requirements, with staffing to be left to the publisher's discretion in craft union jurisdiction.

4 Freedom for the publisher to merge job functions and departments.

5 Confirmation of the publisher's right to subcontract printing and composing work.

6 Conversion of the newspaper's unionised distribution system to an independent contractor system, with displaced employees given first consideration for the contract positions.

7 A review period in the editorial department of up to six months to enable the publisher to determine, at his or her sole discretion, who would be retained.

8 All general forepersons would be required to withdraw from their unions.

9 An express prohibition against sympathy strikes.

10 A reduction of paid holidays from twelve to eight days.

11 Elimination of life-time guarantees for any people who remained on the staff in typographical and stereotyper jurisdiction.

12 Vacations to be scheduled at the convenience of the publisher.

13 The publisher would pay contractual severance to Guild employees who resigned or who were fired over a six-month period, but would

deny such payments to any employee who took a job with a competing newspaper in the Buffalo area within a year. The company would offer termination payments of US$5000 to regular employees on active payroll in the craft unions.

14 Items affecting individual unions would be addressed in separate talks with their leaders (Needham 1989; *Guild Reporter* 1982b).

These demands indicated that before News America would purchase the *Courier* it required agreement from the unions to a total reorganisation of the workplace, involving complete managerial control over staffing levels and job demarcations. Such changes would involve significant initial reductions in staff, to be followed by further cuts over time, and flexibility for management in organising the workplace. Stating the position of News America at the time of these negotiations, Robert Page, a vice-president of the company, remarked: 'We look at a newspaper somewhat differently than many other publishers. We want control of the news room. We want to determine the quality of the staff. We want to put out the type of newspaper we want' (Leapman 1983:250). He also commented that ' "the freedom to select and choose and assess and evaluate" the editorial employees of newspapers was ' 'essential" ' (*Guild Reporter* 1982b:1, 6). In addition, while the fourteen points were put forward as proposals, it soon became evident that they were non-negotiable. The demands made by News America provided an indication of what Page meant when he said the company wanted control of the news room.

The meaning of such demands was apparent also to the unions involved in the negotiations. Typographers and stereotypers realised that they were confronted with the possibility of elimination from the workplace. In meetings of members of the Newspaper Guild, no one suggested that the offer be accepted, and those opposing the company's plans were applauded by the other members present. According to Richard Roth, an international vice-president of the Newspaper Guild and a reporter at the *Courier*, 'the bottom line was that [News America] wanted the ability to fire whomever they wanted. That's a union issue and a journalism issue' (Needham 1989:6). Roth 'called the concept of permitting the News America management to "hand pick" who would be retained "repugnant" ' claiming '[t]hey want people who can think like Rupert Murdoch and who believe in his publishing objectives' (*Guild Reporter* 1982b:6) According to Richard Ramsey, the executive secretary of the Newspaper Guild's contract committee, the selective dismissal proposal put forward by News America would mean that 'the publisher could go through the newsroom and distribute termination notices at will, with no reason necessary and no recourse for the employee' (Needham 1989:7). In addition, employees who left the *Buffalo Courier* under the ownership of News America would 'be unable to keep their contractual severance payment or their US$5000 termination payment if they went to work for a competing newspaper' (Needham 1989:7).

By attempting to augment the scope of managerial prerogative within the workplace, these demands would also have consequences for the introduction of technology into the workplace. By reducing the bargaining power of labour, through reducing the scope of issues over which management would negotiate, the ability of the workforce to influence technological innovation in the workplace in the future would be constrained. These demands were also made possible by the existence of such labour-saving technology. In this particular instance, however, negotiations were unsuccessful from the perspective of News America with the unions refusing their demands, and the last edition of the newspaper was printed and distributed on 19 September 1982. As a consequence, 850 full-time and 250 part-time workers became unemployed with the closure of the newspaper.

In November 1982, News America made a similar offer to acquire the *Boston Herald–American*, as it was called under the ownership of the Hearst corporation. The *Herald–American* had been formed in 1972 as a result of a merger between the *Record American* and the *Herald Traveler*. In 1980, James Doris, who was the publisher of the *Herald–American*, told the union leaders, who represented the 730 employees of the newspaper, that without economic concessions he would recommend that the paper be closed. Among the concessions Doris sought were an indefinite wage and fringe benefit freeze, and a further reduction in the workforce of 200 employees, on top of the 315 that had been removed in 1978. Despite such attempts to reduce the costs associated with the newspaper, by 1982 the *Herald–American* was reported to be losing US$1 million a month. Entering into discussions with Hearst Corporation, News America said it would buy the newspaper only if the eleven unions involved would agree to certain conditions, including the elimination of a number of guild jobs and all job security for employees in the Guild-represented advertising department. In addition, there was to be a six-month period during which management could dismiss any of the advertising employees of the newspaper in any order it might choose, in which cases the dismissed employees would be paid 150 per cent of the contractual severance payment with the payment coming in twelve monthly instalments. These payments would be stopped if the employee took a position with any of six named Boston area newspapers. Further conditions included extension of the unions' contracts for a three-year period, with a 5 per cent wage increase in the first year and cost-of-living adjustments of up to 5 per cent in the final two years; the elimination of a ban on dismissals resulting from the introduction of automated processes; and the imposition of a one-year limit on the time a person dismissed for economic reasons could remain on the rehire list. In total, it was reported that News America was seeking to eliminate one-third of the 830 full-time positions covered by the eleven unions; with craft unions being offered US$20,000 buyouts for each employee (*Guild Reporter* 1982d, 1982e).

In contrast to the negotiations in Buffalo, News America was successful in achieving agreements with the unions at the *Boston Herald*. In particular, the unions agreed to the proposed staffing cuts. Paying the Hearst Corporation US$1 million, plus up to US$7 million for future profits from the newspaper, News America announced that it would invest US$15 million in the newspaper, including US$4 million to pay for job buyouts. However, the Guild was able to resist the demand that there be eliminated from the contract a ban on dismissals resulting from the introduction of automated processes, and was also successful in rejecting the proposed imposition of a blanket one-year limit on the time a person dismissed for economic reasons could remain on the rehire list. On achieving initial workplace reductions, News America then began to bolster its newsroom by introducing staff from its newspapers in Australia and Britain, as it had done at the *New York Post* in the 1970s, while also installing new computer technology, and increasing circulation with a new layout and different news coverage (*Guild Reporter* 1982d, 1982e; Leapman 1983; Maraghy 1989; Tucille 1989).

In 1994, News America sold the *Herald* to Patrick Purcell, as their ownership of the *Herald* and a Boston television station meant the company contravened the same federal cross-ownership law that had forced it to sell the *New York Post*. Patrick Purcell, the new owner of the *Herald*, had been its publisher since 1982, and had also been the publisher of the *New York Post* since 1993. On purchasing the *Herald*, Purcell, who had also been an executive vice-president of News America Publishing Inc., was to claim that '[o]ne thing we do not have any more are the deep pockets and the safety net of Rupert Murdoch's corporation. We are going to have to make it on our own' (*Guild Reporter* 1994a). Given a concern with reducing costs, Purcell soon came into conflict with the newspaper unions at the *Herald* over technological issues, arguing that computer technology made it necessary for compositors and engravers to be made exempt from union representation, and that a new advertisement department should also be exempt from union representation. In response, the unions argued that this was a first step towards the creation of a union-free workplace. Ultimately, this particular dispute was resolved allowing workers in the new department to belong to either the Guild or the CWA/ITU (*Guild Reporter* 1994a, 1994c; 1994f).

The goal of News America in both Boston and Buffalo was to take over a financially struggling newspaper and make it profitable by reorganising workplace relations. A central means by which this was to be achieved was through reducing the influence of unions, thereby gaining control of the workplace. Technology was important in these negotiations – as it had been at the *New York Post* – in that it provided one means of reducing the required workforce while also providing the possibility of introducing an alternative workforce. That the outcomes were different in the two cases can be attributed in large part to the role of the unions. The resistance of the unions in Buffalo was influential as News America did not want to enter a workplace and be confronted immediately by industrial conflict. Perhaps reflecting on the outcome of the *Buffalo Courier* case, the unions in Boston and management at News America were both more flexible in their negotiations with the ultimate result that the paper continued production and the unions were not totally disempowered.

Conclusion

It is apparent that the newspaper industry in the United States has undergone major processes of reorganisation of the workplace in the last twenty to thirty years, and that the development and application of computer technology has been significant in that process. Technology has been linked to drastic job losses in some areas of newspaper production, but also related to the creation of new areas of work. Technology has also been influential in struggles over control of the newspaper workplace resulting in a very different mode of workplace organisation. In particular, unions which were once influential in areas such as staffing levels and work scheduling practices are now increasingly subject to the discretion of management. The evidence presented in this chapter, in combination with the earlier analysis of the institutional context of the US in this time period, suggests, however, that the connections between technology innovation and workplace reorganisation need to be located within the broader societal context. While the practices that have developed in the newspaper industry are bound up with the history of specific workplaces, they are also linked to developments in workplace practice across the United States and to a modified industrial context.

A common feature across industry in the US has been the attempt to use technology as a means of cutting costs and raising profits in a period of economic instability. As a number of commentators have noted, this process has received increasing support from state institutions including the government administration, as well as administrative bodies such as the NLRB and judicial bodies such as the Supreme Court. Responding to economic problems, these state institutions and actors have attempted to create a new context for business in the US. Developments including judicial approval of permanent strike replacement workers, and administrative repeal of some of the foundations of collective bargaining, created a context in which employers and managers could pursue technological innovation in a relatively unfettered fashion. At the workplace itself, renewed aggression on the part of management in its dealings with employees has been witnessed not just in the newspaper industry, but more broadly across US industry. As well as engaging in hostile activities towards trade unions, the practice of providing incentives for contract termination illustrates that contracts will not necessarily extend to their agreed conclusion. Within the newspaper industry itself, the growth of media groups provides another challenge to the processes of technological innovation and workplace reorganisation. The use of similar technology across workplaces, when added to the increasing concentration of ownership, provides employers with an important tool in bargaining with employees. As was evident in the 1978 strike at the *New York Post*, News America was able to bring in employees from its San Antonio operations as a potential alternative workforce.

As in the cases of Australia and Britain, the developments in the US newspaper industry suggest that an explanation of the relationship between technological innovation and workplace reorganisation based on technological determinism is unsatisfactory. The evidence presented in this chapter shows that the influence of technology on workplace organisation varies across workplaces. Even within a single company, such as News America, staffing levels for press operators differed between San Antonio and New York City, and such variation was related to factors such as the presence of unionisation in the workplace. In addition, the different outcomes of the involvement of News America with the *Buffalo Courier* and the *Boston Herald* also indicate that local practice is important. It must also be recognised, however, that News America adopted a confrontational approach in its negotiations with newspaper unions across the US. At each stage of involvement of the company in the newspaper industry it has sought to reduce the influence of unions. Its relative success in this regard is linked both to its willingness to invest in technology and to its ability to draw on alternative sources of labour.

Both the institutional social choice and the labour process models of technological innovation and workplace reorganisation provide more insight into the relationships under examination. It certainly appears that in the case of the newspaper industry in the United States, management and employers have exercised extensive power in bringing about workplace reorganisation, and they have been especially concerned with regaining various managerial prerogatives that had been conceded to unions. Vital in this regard is the issue of determining staffing levels in various departments. It seems evident that in cases ranging from the *Washington Post* in the early 1970s, to the involvement of News America in the New York newspaper strike of the late 1970s, a major concern of management was to re-assert its perceived right to control staffing levels in the production centre and in the news room. To this extent, the emphasis of labour process theory on class relations and on class-based organisations is useful in explaining the events that occurred. The societal context of class relations in US society in the 1970s and 1980s has been reflected in developments in the US newspaper industry, where employers and management have been able to weaken the influence of unions in the workplace.

At the same time, various state institutions were important in the disputes examined in this chapter, while the specific features of both individual newspaper workplaces and individual newspaper workplace relations were also more important than labour process theory would allow. The splintering of both management and union unity in New York was significant to the outcome of the dispute, with the relevant organisations not necessarily acting as might be predicted by class analysis. Choosing to act according to their subjective understanding of the situation supports the proposition of the institutional social choice model that the manner in which new technology is introduced into the workplace is the contingent outcome of historically developed workplace relations. While newspaper unions in Buffalo followed what they considered to be the best course of action by refusing the terms demanded by News America, so did the unions in Boston by accepting many of the terms offered, resulting in different outcomes in the two cases. Nevertheless, it is also the case that even where differences occurred in the relationship between technological innovation and workplace reorganisation, the outcome has been that management control in the workplace across the different cases has intensified, as is predicted by the labour process model.

The outcomes of these various cases indicate the relevance of the relational model of workplace reorganisation. Developments in the newspaper industry in the US suggest that it is important to take seriously contingent circumstances of specific workplaces. While News America was able to purchase the *Boston Herald*, it failed in its attempt to purchase the *Buffalo Courier*. While the circumstances of both cases appeared to be similar, the different practices and perceptions of the unions were fundamental to the variation in outcome. At the same time, the outcome of technological innovation and workplace reorganisation in all the cases considered has been to increase the control of management in the workplace over the workforce, in particular by reducing the role of unions. Such an outcome was also influenced by events across the US from the 1970s to the present, including renewed management hostility to unionisation, and political and legal support for increased employer flexibility in organising the workplace.

As with the English and Australian case studies, the experience of the US shows how important it is to consider not only specific workplace practices and relations but also societal conditions, when seeking to explain processes of technological innovation and workplace reorganisation. The events examined in this chapter occurred primarily before those considered in Australia and England, and as a result, issues of globalisation and of internationalisation were not so immediately relevant. Even in the US, however, the global expansion of News Corporation did have some influence on the relationship between the introduction of new technology and workplace reorganisation. In particular, the potential to draw on a foreign workforce and on foreign resources was always present in its dealings with newspaper unions in the United States. Also, the opportunity to pursue innovation and reorganisation in the United States was to be important for News Corporation when it attempted similar processes in other countries. In the concluding chapter, the factors influencing workplace organisation are considered further, as the evidence from the international cases is compared, and an evaluation is made of the analytic value of the various theoretical approaches that have been presented to examine relationships between the introduction of technological innovation and workplace reorganisation.

CHAPTER 8

Conclusion:
News Corporation: Combining the Global
and the Local

Developments in the newspaper industry from the 1970s to the 1990s provide important insights into the relationship between technological innovation and workplace reorganisation. The analysis presented in this book of the negotiations, disputes and agreements that have been reached at newspapers around the world has illustrated the need to examine not just the workplace itself but the societal context in which workplace reorganisation takes place. By focusing on a company such as News Corporation it also becomes evident that part of this societal context concerns the development of global investment of capital in technological innovation.

The Workplace

The comparative national and global perspective adopted in this study has revealed both similarities and variations in the processes and outcomes of negotiation over technology and workplace reorganisation. All of the newspapers analysed have undertaken extensive programs of technological innovation and workplace reorganisation since 1970, and similar technologies were introduced in each instance. The fundamental technological shift involved the introduction of computerised pre-press production systems to replace hot metal techniques, while there were also computer-related developments in press production systems. Wherever such systems were introduced, a modification in the skills required in the newspaper industry occurred, and the possibility existed of challenging pre-existing craft specialisation and job demarcations. Fundamental differences emerged between sites in their response to such challenges, however, and these differences were related to the processes and outcomes of negotiation between actors in the workplace over innovation. At News International in London the process of innovation was highly volatile and at times violent as management, with the overt support of the government, used the process of technological innovation as a means to remove unions from the workplace. The end result has been a workplace where workers are not represented by unions. While various human resource programs have been established, the basis of bargaining is that workers must respond to the initiatives of management on an individual basis. While the company claims that its workforce is paid well and workplace conditions are excellent, the workforce has little or no involvement in decisions about such organisational matters. Members of the workforce are dependent on management and the proprietors for their well-being. The number of workers employed was also reduced dramatically in certain areas as the Fleet Street production force was largely replaced by newcomers to the industry such as electrical and non-craft workers, and was reduced further with the subsequent removal of job demarcations. The concept of flexibility at News International was a means by which the employer and management attempted to reduce their operating expenses by reducing the costs associated with labour.

The *Financial Times* offers an immediate contrast to developments at News International. While sophisticated new technology has been introduced at the *Financial Times* and significant workplace reorganisation has occurred, unions are still recognised and limited negotiation concerning wages and workplace conditions still occurs between management and unions. A major difference between the two cases is the fact that the Wapping dispute occurred before the *Financial Times* undertook its program of reorganisation. Management at the *Financial Times* were able to take advantage of the developments at Wapping to pursue their own strategy, while unions were concerned to avoid another bitter dispute which could ultimately result in their removal from the workplace.

Murdoch himself acknowledged the influence of News International on developments at other English newspapers, commenting in a 1988 interview that the Wapping dispute had altered perceptions in Britain about what was possible. While Murdoch acknowledged that he had made enemies because of what he had done, he argued that the developments at Wapping had been good both for the newspaper industry and for the country. In particular, he claimed that the power of the unions had been diminished, resulting in greater productivity, while existing newspapers had become financially viable and opportunities had been created for new newspapers to start up (Murdoch 1988). Although management at the *Financial Times* had a reputation for being more receptive to the concerns of labour than management at News International, there seems little doubt that they would have used similar tactics to those employed by News International if the unions had not accepted the terms being offered, which involved major job reductions and a fundamental reorganisation of job demarcations.

Developments at the *Advertiser* in Adelaide also occurred in the shadow of the Wapping dispute, although in this instance it appears that neither management nor labour was prepared or willing to enter into such a dispute. While there had been tensions in labour relations at the *Advertiser*, these were not of the magnitude of those that had historically existed in Fleet Street. At the *Advertiser*, the unions had been more willing over time to accommodate technological innovation into the workplace, and had successfully retained various demarcations and controls in the workplace. Nevertheless, unions and management involved in workplace negotiations at the *Advertiser* were aware of the dispute at Wapping, and events in the early 1990s suggested that confrontation was perhaps becoming a part of workplace relations. In addition, the progressive introduction of technology into the workplace was reducing rapidly the number of jobs in the pre-press area, although the unions were still having some influence in negotiating about the influence of such technology.

Developments at newspapers in the United States, such as those owned by News America, also had severe repercussions for the unions and the workforce. While the role of unions as representatives of workers in the newspapers of News America has been challenged by the company, they have so far avoided a repetition of the outcome at Wapping. Nevertheless, management and unions at US newspapers have been involved in a series of buybacks influenced by technology in which unions cede various workplace controls in return for job guarantees. Despite such job guarantees, the effect of these developments related to technological development has been devastating for the workforce, with staff reductions in the pre-press area as drastic as in Adelaide and London.

The outcomes of these cases of technological innovation suggest that local practice is fundamental to the development of workplace organisation in the newspaper industry. The variation in local relations between management and unions has been one factor that has influenced the relationship between technological innovation and workplace reforms. Part of the reason for variations in workplace practice at individual newspapers was explained by Murdoch, who in a 1993 interview claimed that on a daily basis it was up to the executives at each newspaper to manage the business and to solve problems as they arose. Murdoch claimed that he would become a more visible presence in crisis situations (Coleridge 1993a, 1993b). Murdoch has also stated that News Corporation does not have a single culture, but rather that styles of management across the company range from autocratic to democratic (Murdoch 1988).

The case studies provide support for these descriptions. Management in London and New York adopted a confrontational approach to their interactions with the workforce, for example, which has not been evident to such an extent in Adelaide. Such variations are linked, in part, to the historic development of workplace relations between management and unions in the workplace. In England and the United States the unions were perceived by management to be obstructive and unresponsive to technological and organisational developments that management believed were crucial for the financial future and competitiveness of the newspapers. In contrast, the unions at the *Advertiser* were perceived by management to be more willing to enter into negotiations with management under the prior ownership, and this continued at least in the early years of ownership by News International.

Despite the negotiations with unions at the *Advertiser*, Murdoch has indicated his understanding of the relation of the company to unions and the workforce. In regard to the general role and function of unions, Murdoch has argued that the role of unions is declining, especially as people become more flexible in their careers, and less fearful of losing specific jobs. Murdoch has stated further that he is opposed to union representatives being involved on management boards as they have outside interests to represent. By contrast, he does see a role for employee representatives being involved in the managerial process. Allowing employees, through representatives, to be involved in such processes is one way of allowing them to feel a sense of belonging to the organisation, and this in turn will encourage them to maximise their efforts for the company (Murdoch 1988).

The evidence presented in this book suggests that while unions are less involved in negotiations in the workplace in the 1990s than they were prior to the introduction of computer technology, this is related more to the outcome of struggles between management and unions than to the development of new beliefs on the part of the workforce. Unemployment is still a major concern for workers in the industry, and at workplaces such as the *Advertiser* there has been active resistance to and negotiation over the removal of demarcations between jobs. The 1993 national newspaper strike in Australia was one indication that the concept of 'changing jobs' is not always acceptable to union workers, especially if the new jobs created are to be filled by members of another union or by non-union workers. In addition, it is not evident that workers in the industry are less inclined to seek union organisation and representation. Recent developments at News International indicate that the workforce is seeking ways of becoming organised in a workplace where unions have been banned. Even in Adelaide, there has been a recent failed attempt by management to encourage some of the workforce away from union membership. Rather than the lessening role of unions being attributable to varying perceptions of the role of organised workplace representation, it can more appropriately be attributed to the outcome of workplace disputes which altered fundamentally the balance of power in the workplace.

The Institutional and Societal Context

It has been emphasised throughout this study that developments at the workplace cannot be understood in isolation from their institutional and societal context. By analysing the operations of a global company in various national locations, the importance of this claim in explaining workplace developments has become evident. Each of the countries in which News Corporation operated, as examined in this study, was undergoing major restructuring between the 1970s and the 1990s. Developments, including relatively high levels of unemployment, falling or stagnating gross domestic products, movement of employment from manufacturing to services, the reorganisation of corporate structures, and the rapid development and diffusion of information and communication technologies, demanded responses from national actors, including the state, which were to have ramifications at the workplace.

In both Britain and the US, many of the processes I have examined occurred in a situation where a conservative government was in office, and was attempting to create a context in which business was able to invest and operate with few constraints. From the early days of both the Thatcher government and Reagan administration, trade unions were identified as a fundamental constraint on business activity, including investment in technology. In both countries, much of the legislative and policy activity of the governments focused on removing or diminishing the influence of unions in the workplace, and in pursuing this goal the governments were assisted by decisions of judicial and administrative institutions. Such antagonism towards unions, and overt support for business, was evident in disputes in the newspaper industry. Close personal links between Murdoch and Thatcher in Britain were evident during the Wapping dispute where the government strongly supported the tactics and goals of News International, while condemning the unions. While the personal links between Murdoch and Reagan in the US were not so obvious, the government and various state institutions had taken steps which were important in allowing newspaper proprietors to confront unions over technological innovation and workplace reorganisation. Of crucial importance were enactments concerning the use of permanent strike replacements, and confrontations between the administration and various unions including the Professional Air Traffic Controllers Organization (PATCO), while the National Labor Relations Board (NLRB) removed many protections won by unions in earlier struggles.

While the Labor Party government in Australia was responding to similar global circumstances, and was concerned to create a context in which business would invest to raise productivity, the approach of the government was to emphasise consensus and planning. While reorganisation of industrial relations occurred, and the government supported moves by the union movement towards rationalisation, agreements such as the Accord indicated that the unions were considered to be crucial partners in the future development of Australian society. The Accord process, and the related development by the unions of a policy of strategic unionism, were reflected in the *Termination, Change, and Redundancy* (TCR 1984) case held before the Commonwealth Conciliation and Arbitration Commission, and in events in the newspaper industry. Throughout the 1980s, the unions at the *Advertiser* attempted to adopt a unified approach to negotiations about technological innovation, while management also sought to achieve innovation and workplace reorganisation through negotiated agreement. As in Britain and the United States, however, in most instances management held the initiative in these processes, while unions and the workforce were placed in the position of responding to the initiatives of management. The relationship between management and workers was to be exacerbated as the issue of the global holdings of News Corporation became increasingly relevant.

The Globalisation of the Workplace

Beginning in the 1970s and becoming increasingly influential in the 1980s was the development of global relations which were to affect actions at the workplace. While there are practices and relations specific to the individual workplace, this investigation has shown that there have also been global developments which have affected the newspaper industry and which are related both to technology and to ownership. There are close links between the types of technology being used at Wapping in London and in Adelaide at the *Advertiser*, and at other Australian newspapers. Indeed, the technological developments relating to factors such as colour print in Australia were modelled initially on the progress made at Wapping, and when the experiments in Australia proved successful it was anticipated that these further developments would be introduced in Wapping. Another global development of importance for the workplace has been the continuing expansion of News Corporation. Not only has this expansion allowed for the sharing of technological resources by various publications of the company, it has also allowed the company to reduce its labour force requirements. As newspaper companies such as News Corporation have extended their ownership around the world, even potential beneficiaries of the capabilities of computerised technology including journalists have been threatened by job loss, as stories are used in more than one newspaper, and as increasing use is made of syndicated stories and news services. The possibility of sharing resources within and between countries has also been important to News Corporation in its dealings with its workforce. As the potential for a strike developed at the *New York Post* in the late 1970s, for example, News America was able to bring in its workforce from its San Antonio newspaper so that production could continue even if a strike eventuated. The use of such tactics is also clearly possible in both Australia and Britain where News Corporation has more substantial newspaper holdings. While it is less immediately evident that any single mode of workplace organisation is developing in the holdings of News Corporation, there are some indications of similarities developing across geographically distinct workplaces based on the Wapping model. In its US newspapers, and increasingly in Australian newspapers such as the *Advertiser*, attempts have been made by management to gain complete control over the workplace so it is able to remove demarcations altogether, while it also appears that the company is attempting to introduce individual contracts of employment. Such developments would then make it more straightforward for the company to exert control in the workplace to the extent that it has done so in London.

There are local, national and global processes at work in the newspaper industry, but the effects of such processes on the workforce are quite uneven. Global dispersion of technology is posing a major challenge to print and craft workers in the industry, while also providing potential opportunities for white collar and professional staff. The continued growth of companies such as News Corporation means, however, that even the potential beneficiaries of technological development in the workplace may be under threat as companies seek to reduce costs by combining resources such as labour. In the following section, the theoretical models that have been proposed throughout this study to explain such developments are considered, and a modified relational model is suggested.

Theoretical Implications of the Newspaper Industry

An important part of this book has been the examination of various theoretical approaches to technological innovation and workplace reorganisation. The lessons learned from the newspaper industry concerning the explanatory value of each of the models considered here have important implications for analyses both of future developments in the newspaper industry and for examining relationships between technological innovation and workplace reorganisation in other industries. The first model to be examined was technological determinism which proposes that:

1 new technology directly influences workplace organisation; and
2 this relation will in part be mediated by patterns of management and
· by relations between management and labour.

The cases analysed in this study suggest that technological determinism is inadequate as an explanation of technological innovation and workplace reorganisation. Certainly, technology does influence workplace relations. In all the newspapers studied, pre-press and production work have been dramatically affected by computer technology, while the scope of work available to journalists is potentially much broader. In addition, the technology adopted by the various newspapers is in a broad sense quite similar. Nevertheless, the crucial question of how work and workplace organisation are precisely influenced by technology is not answered simply by considering the type of technology being introduced. While similar technology has been adopted by newspapers in Wapping, New York and Adelaide, the outcomes for the workforce have been quite different. Ranging from issues of the continued unionisation of the workplace to the existence of demarcations between jobs, workplace and societal relations must be analysed to examine more precisely the influence of technology on the workplace. It is important also to investigate the context which affects the introduction of technology into the workplace. A motivation of News International in introducing technology into its London newspapers by moving to Wapping was to remove the influence of unions from the workplace, and the company was assisted in achieving this goal by the Conservative government. While a similar goal was apparent in developments in US newspapers, no such overt intention concerning unions was evident at either the *Advertiser* (at least until recently) or the *Financial Times*. Technological determinism indicates the potentially important relation between new technology and patterns of management–worker relationships. The model fails, however, to take into account the complexity of factors within organisations that mediate the relationship between new technology and workplace reorganisation and the external factors which shape the relationship. That is, what needs to be examined more carefully are the historically developed relations at specific workplaces, as influenced by economic, political and social conditions.

One response to the shortcomings of the approach of technological determinism has been the development of labour process theory which proposes that:

1 the process of technological innovation and subsequent workplace restructuring is shaped primarily by the location of the particular workplace within a specific set of class relations;
2 although management ultimately controls the process of introducing technology and reorganising the workplace in a capitalist economy, the workforce and its representatives play an active role in the process by either resisting or accommodating technological innovation on the shopfloor;
3 at a particular moment in the development of class relations, a particular set of workplace relations tends to predominate across industry; and
4 the relationship between the introduction of technology and subsequent workplace reorganisation can, in general terms, be predicted from one workplace to the next, based on the stage of development of capitalism.

It appears from all the cases examined that the introduction of technology into the newspaper workplace was influenced to some extent by the interaction between class-based organisations, and that class relations themselves were important in the processes. In the United States the interactions between class-based organisations, including unions and management, were vital to outcomes in Boston and Buffalo, while in New York, class relations between management and unions, and intra-class relations between unions in the New York industry, were central to workplace reorganisation. In New York, management was able to exploit inter-union rivalries effectively, as various unions resisted technological innovation while others were more compliant with the position of management. Resistance or compliance to technological innovation was also associated in some instances with the position of the organised workforce in the labour process. While print workers resisted job-threatening technology, journalists were less resistant to technology which did not appear to pose an immediate threat to their position in the workforce. Similar examples were apparent in the relationships between technological innovation and workplace reorganisation in newspapers in Britain and Australia.

Management control in the workplace, as emphasised by labour process theorists, is also critical to the newspaper industry. Events occurred which support the claim of Burawoy that the 1980s was a period in which there was a shift from hegemonic forms of factory production, where consent was a dominant mode of shopfloor inter-action, to a form of hegemonic despotism where capital is using its resources to extract concessions from labour (Burawoy 1985). In devel-opments at the *Washington Post* and the *New York Post*, for example, the introduction of computerised technology was used explicitly by management as part of a strategy to gain control of the production room. While the immediate union response was to take industrial action, the shift in the balance of power in workplace bargaining to management was made apparent when the unions eventually agreed to cede control in exchange for job guarantees for members. Such a concession on the part of the unions would have been unthinkable only a few years earlier, and provides an indication of the influence of the interaction of a number of factors, including the availability of computer technology, relatively high levels of unemployment and concentration of ownership in the industry.

Perhaps the major shortcoming of labour process theory involves the explanation of variations that occur in the relationship between tech-nological innovation and workplace reorganisation in different firms within the same industry. More emphasis needs to be directed at exam-ining how organised workers continue to struggle against exercises of managerial power in differing workplaces. While events surrounding technological innovation and workplace reorganisation at the *Financial Times* were certainly influenced by the Wapping dispute, there were important differences in the resulting workplaces. The continued exis-tence of unions at the *Financial Times* was particularly significant. While the current balance of workplace bargaining power means that these unions are not as influential as in previous times, there exists the poten-tial for collective bargaining at the workplace. In contrast, the removal of unions from the workplace at Wapping means that workers there are confronted with a much more difficult situation in terms of making demands on management, either at present or in the future.

A model that is potentially more sensitive to local workplace practice, and which permits a more thorough analysis of variations in the outcomes of technological innovation on workplace reform, is the insti-tutional social choice model which proposes that:

1 in considering external processes which influence technological innovation and workplace reorganisation, the role of the state is fundamental. State policies and administrative decisions, legislation

and judicial rulings, and the struggles between politicians and political parties, all influence the context in which innovation and reorganisation occur;

2 the influence of social and economic relations on technological innovation and workplace reorganisation is mediated by these state institutions and political actors;

3 the manner in which new technology is introduced into the workplace, and the associated reorganisation of workplace relations, are the contingent outcome of historically developed workplace relations. The development of these workplace relations is based on interactions between management and workers, management and unions, and unions and workers;

4 while it is possible to identify in advance the actors who need to be analysed, including management, unions and workers, and the state, it is not possible to predict the behaviour and interests of these actors in specific situations in advance. This includes the interests of actors who may appear to share a similar objective situation, but in fact behave differently in specific situations; and

5 analysis of the relation between technology and workplace reorganisation should proceed on a case-by-case basis, focusing on the specific circumstances of each workplace. There is no necessary correlation between events at one workplace and at another workplace, and generalisations across workplaces should not be made.

The emphasis on state institutions and policies deemed essential by the institutional social choice model is supported by the various case studies presented in this book. In each situation, the strategies adopted by management and the responses of labour were influenced in important and significant ways by the state. In Britain, for example, it appears evident that there were strong links between the development of the plans for Wapping and government legislation related to factors such as secondary picketing and strike action. In addition, the support for the actions of the company by leaders within the government, while not decisive, acted as a signal that their activities were justified, and that the actions of the unions were not similarly justified. Similar instances of the importance of state institutions and state actors in understanding relations between innovations and reorganisation were also documented in the cases from Australia and the United States.

The emphasis of the institutional social choice model on analysing specific cases, and not generalising in advance about the interests of actors at the workplace, also provides insights into developments such as those which occurred in London, and in Buffalo and Boston. Using a strict class analysis, it might have been expected that management at the *Financial Times* would have confronted the print unions in the same manner as occurred with News International. The specific relations at the *Financial Times*, and the influence of the Wapping dispute on the world view of the workplace actors meant, however, that the interests of management and unions were defined in different terms. Management at the *Financial Times* could accept the continued representation of the workforce by unions, aware that they would still be able to exercise a high level of control in the workplace.

Variations also occurred in developments in the US where, confronted with very similar situations, the unions in Buffalo resisted the takeover attempt by News America, resulting in the closure of the newspaper. In contrast, the unions in Boston chose to enter into an agreement with the company. Whereas it would be difficult to explain these differing outcomes purely in terms of class interest, the focus of the institutional social choice model on subjective interests and specific circumstances in firms provides a possible explanation. The unions in Buffalo resisted demands which they believed conceded too much to management, while News America was not prepared to enter into negotiations over its conditions. Having witnessed these events, the specific circumstances of the situation in Boston meant that both sides were more willing to enter into negotiations. In both of these examples, the outcomes are explicable only by analysing the specific circumstances in detail, and by recognising that there may be variations in the interests of class-based actors across time and location.

It is difficult to ignore, however, the common outcome across all the cases that developments in the newspaper industry resulted in an intensification of the control of management in the workplace, as predicted by labour process theory. While there are variations in the extent of such intensification, there appears to be a tendency developing towards a particular outcome. The development of such a relationship between new technology and workplace outcomes suggests the relevance of the relational model of workplace reorganisation which was derived from both labour process theory and the institutional social choice model, and proposes that:

1 while employers and management control the initial introduction of technology into the workplace through both their access to financial resources and the structural conditions of capitalism as an economic system, the workforce and its representatives influence that process through their capacity to either cooperate with or resist the introduction of technology into the workplace;
2 the institutions of the state – including the legislature, courts, and various departments or ministries – and actors associated with the

state, including political parties, influence the process of technological innovation through various processes whether intended to facilitate or constrain the activities of employers, management and unions. These processes may include legislation, legal decisions, and explicit or implicit policy;

3 workplace reorganisation associated with the introduction of new technology occurs through interactions between these same actors and institutions, although the introduction of new technology into the workplace may influence the balance of power between those actors and institutions. That is, the introduction of technology and workplace reorganisation will be influenced by the historically developed and developing relations between management, labour and the state; and

4 these historically developed relations are shaped and constrained by their location within a particular economic, political and social formation, and the processes and outcomes of technological development and workplace reorganisation are similarly shaped and constrained.

As indicated, the relational model attempts to combine elements of both the labour process model and the institutional social choice model to explain the relations between technological innovation and workplace reorganisation. While the case studies presented in this study have indicated support for the relational model, they also suggest that there is a need for an explicit awareness of the global dimension of an industry, especially in regard to the dispersion of technology. In Figure 8.1, a modified relational model is presented which attempts to illustrate the relationship between industry-wide changes and local firm reorganisation.

The modified model proposes that the global dispersion of technology has a direct influence on the technology used in local workplaces. However, the relationship between industry-wide workplace reform and local workplace reorganisation is mediated by management–worker relations in specific firms and influenced by the context in which those local firms operate. It is possible, of course, that later experiences in specific firms have a reciprocal influence on the evolution of industry-wide reforms. Although such frameworks present a necessarily restricted picture of the complexity of interacting influences, the modified model is meant to represent a dynamic set of effects that need to be examined for a more complete understanding of variations in workplace relations.

The cases of News International, the *Financial Times*, and the Adelaide *Advertiser* suggest that there is a process of global integration occurring whereby the practices of one firm within News Corporation in regard to technological innovation and workplace reorganisation influences other firms within the industry also engaged in innovation and

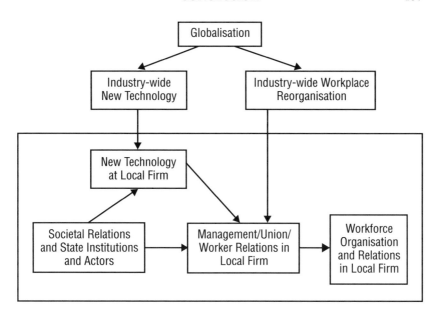

Figure 8.1 Modified Relational Model

reorganisation. Developments at the *Financial Times* and at the *Advertiser* for both management and unions, for example, were influenced in important ways by the events at Wapping. The unions at the *Financial Times* and at the *Advertiser* were well aware of the consequences of Wapping for their union colleagues and were not keen to repeat such an experience. Especially at the *Financial Times*, management was able to use the tactics employed by News International as a bargaining tool in its interactions with its own workforce. While the company was able to claim that it would negotiate with the workforce, the possibility of repeating the strategy used at Wapping was made very clear. The situation at the *Advertiser* was somewhat different, in that innovation and reorganisation were occurring in a different institutional and societal context. Nevertheless, the Wapping experience was present in negotiations concerning the introduction of a new production centre at Adelaide.

The experience of Wapping had an influence on workplace relations in New York, with Rupert Murdoch comparing the role of unions at the *Post* to that of the unions in Fleet Street prior to the move to Wapping. Such comparisons were made to highlight differences, but also to suggest that if the activities of the unions resembled too closely those of the Fleet Street unions, then the New York unions were inviting a Wapping situation. The decimation of the Newspaper Guild after taking strike action in the 1990s indicated that Murdoch was true to his word.

The developments considered here do not constitute an end point in the influence of technological development on the workplace and on workplace relations. What has been analysed is the influence of computer technology at the moment when it was introduced into various newspapers. Through the activities of organisations such as News Corporation, there has since occurred a globalisation of effects related to technology which are beginning to influence the organisation of the workplace. The global dispersal of technology means that workers and managements around the world are faced with similar challenges, although the mediation of such challenges through societal, institutional and workplace contexts means that the effects to date in specific workplaces have been uneven.

The Newspaper Industry and Social Organisation of the Economy

Questions were raised in chapter one about the social organisation of capitalist societies, and whether workplace, political and economic relations in such societies are currently undergoing modification or more fundamental reorganisation. The evidence presented in this book indicates that there are developments occurring in the workplace which have important repercussions for society. In particular, a shift has occurred from craft- to computer-based labour in the newspaper industry. Also, there is a tendency developing towards individual forms of bargaining over workplace terms and conditions – if such bargaining occurs at all. At the newspapers of News International in Britain, for example, there is little bargaining between management and unions at least in so far as that concept was understood in the 1970s. The analysis in the current study revealed that terms presented to the workforce by management are in effect non-negotiable. While there may be discussion over issues such as health and safety, management is now in a much more powerful position than it was twenty years ago. A similar situation is occurring in the United States, with employers using technology as one means of circumventing or diminishing the influence of unions in the workplace. In addition, as in the rest of US industry, many newspaper proprietors now operate non-union workplaces. The situation in Australia has been somewhat different, with the unionised workforce receiving support and protection from an arbitration system and more recently from a government committed to negotiated bargaining as a means of improving workplace productivity. Recent events in Australia suggest, however, that this situation is under threat and that unions may be losing some of their influence in the workplace. Intensification of the move to enterprise bargaining, which resulted in the nation-wide newspaper strike in 1993, and the 1996 election of the

Liberal–National Party coalition to government represent important challenges to collective representation in the workplace. In all the countries under analysis, there is a trend towards individual negotiation between management and workers over terms and conditions of employment which represents a significant development away from collective bargaining, particularly in an industry such as newspapers which has historically been heavily unionised.

For those involved in the debate over the social organisation of capitalist economies, the question that arises is whether such developments are empowering for those in the workplace or whether they constitute more correctly an increase in the control that management exerts in the workplace. In the case of the newspaper industry, the latter interpretation appears more appropriate. Even though technological developments have given journalists more scope for activity in the workplace, and have created jobs for electricians and computer workers, in many instances these gains are tenuous. In the United States, concerns about competitiveness and profitability are driving many employers to cut journalistic staff, while strikes at newspapers in cities such as San Francisco and Detroit have been rearguard activities on the part of employees threatened with job loss (*Guild Reporter* 1994f, 1994g). Similar situations have arisen in both London and Adelaide, and in all three countries the global reach of News Corporation constitutes an additional challenge to the workforce. Developments in national political and economic practice are also assisting in this process, with government policies supporting ideas of marketplace freedom for business which has led to policies that remove previous protections for organised labour on the basis that such protections constitute a hindrance to the operation of the market economy. The potential globalisation of workplace relations constitutes a further challenge for organised labour as corporations such as News Corporation may seek increasingly to impose common conditions on their geographically disparate holdings. In other words, this study has indicated that social relations in newspaper organisations mediate the relationship between the introduction of new technologies and the reform of workplace practices and relations. Also, the investigation revealed that these relationships were shaped by state institutions; economic, political and social relations; and by the globalisation of technology in the industry. The analysis of the newspaper industry suggests, therefore, the importance of analysing the connections between events occurring locally, nationally and globally if relationships between technological innovation and workplace reorganisation in industry are to be understood. As international companies continue to expand their global holdings the need for such interconnected analyses becomes even more urgent.

Bibliography

ACTU/TDC (Australian Council of Trade Unions/Trade Development Council) Mission to Western Europe 1987, *Australia Reconstructed. A Report by the Mission Members to the ACTU and the TDC*, Canberra: Australian Government Publishing Service.

Advertiser Newspapers Limited 1978, 1979, 1985, 1986, *Annual Report and Balance Sheet, December 31, Annually, including Notice of Annual Meeting*, Adelaide: Griffin.

AHRSC (Australian House of Representatives Select Committee Report on the Print Media) 1992, *News and Fair Facts. The Australian Print Media Industry*, Canberra: Australian Government Publishing Service.

Alliance 1993a, 'Print Unions Launch Joint Accord Battle', June–August:5.

—— 1993b, 'Alliance Seen as Top 20 Union Material', Summer:2.

—— 1993c, 'News Bosses Try Bullying Tactics', September–December:3.

—— 1993d, 'Wage Rises Follow Historic Stoppage', September–December:3.

—— 1994a, 'News Ltd Staff Strike a Deal', Spring:19.

—— 1994b, 'Combined Unions Lodge Super Log on News Ltd', Summer:14.

—— 1994c, 'Lock-out at News in Adelaide', Summer:2.

ALLR (*Australian Labour Law Reporter*) 1992, *Genesis of Enterprise Bargaining*, Sydney: CCH Australia Limited.

ALP/ACTU (Australian Labor Party/Australian Council of Trade Unions) 1983, *Statement of Accord by the Australian Labor Party and the Australian Council of Trade Unions Regarding Economic Policy*, Melbourne: ACTU.

AMWU (Australian Manufacturing Workers Union) 1998, World Wide Web Home Page: www.amwu.asn.au.

Archer, Robin 1992, 'The Unexpected Emergence of Australian Corporatism', in *Social Corporatism: A Superior Economic System?*, eds Jukka Pekkarinen, Matti Pohjola and Bob Rowthorn, Oxford: Clarendon.

Armstrong, Philip, Glyn, Andrew and Harrison, John 1984, *Capitalism Since World War II. The Making and Breakup of the Great Boom*, London: Fontana.

Arup, Christopher 1991, 'Labour Law, Production Strategies and Industrial Relations', *Law in Context* 9:36–69.

ASTC (Australian Science and Technology Council) 1990, *Science, Technology, and Australia's Future*, Canberra: Australian Government Publishing Service.

Atkin, Michael 1992a, 'The New Face of the *Advertiser* . . . and the People Behind it', *Graphix* 205 May:19–20.

—— 1992b, 'New Year to Herald Colourful Dawn for Daily', *Graphix* 212 December:7.

Auletta, Ken 1997, *The Highwaymen: Warriors on the Information Superhighway*, New York: Random House.

Bagdikian, Ben 1995, *Double Vision. Reflections on My Heritage, Life, and Profession*, Boston: Beacon Press.

Bamber, Greg J. and Lansbury, Russell D. (eds) 1989, *New Technology: International Perspectives on Human Resources and Industrial Relations*, Sydney: Allen & Unwin.

Barlow, Frank 1986, 'Why We Have To Change', *The Financial Times and St Clements Press 1986–1988*, London: Financial Times/St Clements Press.

Bell, Stephen 1992a, 'Business, Government and the Challenge of Structural Economic Adjustment', in *Business–Government Relations in Australia*, eds Stephen Bell and John Wanna, Sydney: Harcourt Brace Jovanovich.

—— 1992b, 'Structural Power in the Manufacturing Sector: The Political Economy of Competitiveness and Investment', in *Business–Government Relations in Australia*, eds Stephen Bell and John Wanna, Sydney: Harcourt Brace Jovanovich.

Bell, Stephen and Wanna, John (eds) 1992, *Business–Government Relations in Australia*, Sydney: Harcourt Brace Jovanovich.

Bellace, Janice 1992, 'The United States', *Bulletin of Comparative Labor Law* 23: 241–53.

Bennett, E.C. 1980a, 'Facing the 1980s', *Printing Trades Journal* LXIV(1) January:1.

—— 1980b, 'Today is the Tomorrow you were Thinking of Yesterday', *Printing Trades Journal* LXIV(5) May–June:1.

Berger, Suzanne and Dore, Ronald (eds) 1996, *National Diversity and Global Capitalism*, Ithaca, NY: Cornell University Press.

Best, Michael 1990, *The New Competition. Institutions of Industrial Restructuring*, Cambridge, MA: Harvard University Press.

Bingel, Joe 1978, 'TNG Convention Hears Speech by President Perlik', *Typographical Journal* 173(2) August:4, 14–15.

—— 1982, 'Reformation Caused by Modern Age of Printing', *Typographical Journal* 180(6) June:3.

Bingel, Joe, Kopeck, Thomas W., McMichen, Robert S., Wartinger, Robert L. and Heritage, Allan J. 1978, 'Battles are Bringing Unions Together', *Typographical Journal* 173(4) October:4–5.

Block, Fred 1990, *Postindustrial Possibilities: A Critique of Economic Reasoning*, Berkeley: University of California Press.

Bluestone, Barry and Bluestone, Irving 1992, *Negotiating the Future: A Labor Perspective on American Business*, New York: Basic Books.

Bowles, Samuel, Gordon, David M. and Weisskopf, Thomas E. 1990, *After the Wasteland: A Democratic Economics for the Year 2000*, Armonk, New York: M.E. Sharpe.

Boyer, Kenneth D. 1991, 'The Reagan Regulatory Regime: Comment', in *The Economic Legacy of the Reagan Years: Euphoria or Chaos?*, eds Anandi Sahu and Ronald L. Tracy, New York: Praeger.

Boyer, Robert and Drache, Daniel (eds) 1996, *States Against Markets: The Limits of Globalisation*, New York: Routledge.

Braverman, Harry 1974, *Labor and Monopoly Capital: The Degradation of Work in the Twentieth Century*, New York: Monthly Review Press.

Brooks, Brian 1992, 'Australia', *Bulletin of Comparative Labor Relations* 23:35–53.

Brown, Gordon 1988, 'Dependency Culture: Welfare for the Wealthy', *New Statesman*, 11 March:10–1.

Burawoy, Michael 1979, *Manufacturing Consent: Changes in the Labor Process under Monopoly Capitalism*, Chicago: University of Chicago Press.

—— 1985, *The Politics of Production. Factory Regimes Under Capitalism and Socialism*, London: Verso.

Bureau of National Affairs Inc. various years, *Basic Patterns in Union Contracts, By the Editors of Collective Bargaining Negotiations and Contracts*, Washington DC: The Bureau of National Affairs, Inc.

Cahill, John 1987, 'Foreword', *PKIU Technology Guidebook*, Sydney: PKIU, 1.

—— (ed.) 1993, *Australian Printing Industry Yearbook 1993*, Sydney: PKIU.

Callus, Ron, Morehead, Alison, Cully, Mark and Buchanan, John 1991, *Industrial Relations at Work. The Australian Workplace Industrial Relations Survey*, Canberra: Australian Government Publishing Service.

Campbell, John and Lindberg, Leon N. 1990, 'Property Rights and the Organization of Economic Activity by the State', *American Sociological Review* 55 October:634–47.

Capling, Ann, Considine, Mark and Crozier, Michael 1998, *Australian Politics in the Global Era*, Melbourne: Addison Wesley Longman.

Carney, Shaun 1988, *Australia in Accord. Politics and Industrial Relations Under the Hawke Government*, Melbourne: Sun Macmillan.

Carroll, William K. 1990, 'Restructuring Capital, Reorganising Consent: Gramsci, Political Economy, and Canada', *Canadian Review of Sociology and Anthropology* 27(3):390–416.

Castells, Manuel 1996, *The Rise of the Network Society*, Cambridge, MA: Blackwell Publishers.

Clark, Jon, McLoughlin, Ian, Rose, Howard and King, Robin 1988, *The Process of Technological Change. New Technology and Social Choice in the Workplace*, Cambridge, UK: Cambridge University Press.

Clement, Ian 1992, 'The EETPU', *The Independent* March 13:15.

Cockburn, Cynthia 1983, *Brothers: Male Dominance and Technological Change*, London: Pluto.

—— 1991, *Brothers: Male Dominance and Technological Change*, 2nd edn, London: Pluto.

Coleridge, Nicholas 1993a, *Paper Tycoons: The Latest, Greatest Newspaper Tycoons and How They Won the World*, London: Heinemann.

—— 1993b, 'The Daily Terror', *The Bulletin* June 8:28–34.

Conley, David 1997, *The Daily Miracle: An Introduction to Journalism*, Melbourne: Oxford University Press.

Cook, Peter 1992, *Best and Fairest. The Quiet Revolution in our Workplaces*, Canberra: Australian Government Publishing Service.

Cornfield, Daniel B. 1992, 'Technological Change and Labor Relations in the United States', in *Technological Change and Labor Relations*, ed. Muneto Ozaki, Geneva: International Labor Office.

Cox, Archibald, Bok, Derek, Curtis, Gorman, Robert A. and Finkin, Matthew W. 1991, *Cases and Materials on Labor Law*, 11th edn, Westbury, New York: The Foundation Press.

Creighton, Breen and Stewart, Andrew 1990, *Labour Law: An Introduction*, Annandale, New South Wales: Federation Press.

Crouch, Colin and Streeck, Wolfgang (eds) 1997, *Political Economy of Modern Capitalism: Mapping Convergence and Divergence*, London: Sage.

Davis, Edward M. and Lansbury, Russell D. 1989, 'Worker Participation in Decisions on Technological Change in Australia', in *New Technology. International Perspectives on Human Resources and Industrial Relations*, eds Greg J. Bamber and Russell D. Lansbury, Sydney: Allen & Unwin.

Davis, Mike 1986, *Prisoners of the American Dream: Politics and Economy in the History of the US Working Class*, London: Verso.

Dean, Brenda 1985, 'New Technology – We Will Defend Our Members', *SOGAT Journal* July:2–3.

—— 1986a, 'Brenda Dean: Frankly Speaking. Frances Morrell Interviews Brenda Dean, General Secretary of SOGAT', *New Socialist* 36 March:28–30.

—— 1986b, 'Report of SOGAT '82 Meeting', *SOGAT Post* February:6–8.

Deb, Karen 1994, 'Tours Keep Alan in Touch', *The Insider* 18 July–September:1.

Derian, Jean-Claude 1990, *America's Struggle for Leadership in Technology*, trans. Severen Schaeffer, Cambridge, MA: MIT Press.

Deutsch, Steven 1993, 'Recent Developments in the US: Trade Union Strategies', *Economic and Industrial Democracy* 14(3) August:327–31.

Deutsch, Steven and Schurman, Susan 1993, 'Labor Initiatives for Worker Participation and Quality of Working Life', *Economic and Industrial Democracy* 14(3) August:345–54.

Donnelly, Walter 1987, 'Letter: Proud to be There', *SOGAT Journal* March:10.

Dorey, Peter 1993, 'One Step at a Time: The Conservative Government's Approach to the Reform of Industrial Relations Since 1979', *The Political Quarterly* 64(1) January–March:24–36.

Dubbins, Tony 1983, 'Cooperation . . . the Key to a Formula for Our Future', *Print* 20(12) December:8–9.

—— 1986a, 'A Message From the General Secretary of the NGA', *London SOGAT Post* February:3.

—— 1986b, 'Solidarity is Also OK at Wapping', *Print* 23(4) April:4.

—— 1987, 'National Council End NI Dispute', *Print* 25(3) March:1.

Edwards, Paul, Hall, Mark, Hyman, Richard, Marginson, Paul, Sisson, Keith, Waddington, Jeremy and Winchester, David 1992, 'Still Muddling Through', in *Industrial Relations in the New Europe*, eds Anthony Ferner and Richard Hyman, Oxford:Blackwell.

Eisen, David J. 1978, 'TNG, ITU – From Pen Pals to Going Steady', *Typographical Journal* 173(1) July:39–48.

Elgar, Jane and Simpson, Bob 1993, 'The Impact of the Law on Industrial Disputes in the 1980s', in *New Perspectives on Industrial Disputes*, eds David Metcalf and Simon Milner, London:Routledge.

Emery, Michael, Emery, Edwin with Roberts, Nancy L. 1996, *The Press and America: An Interpretive History of the Mass Media*, 8th edn, Boston: Allyn and Bacon.

Evatt Foundation 1995, *Unions 2001: A Blueprint for Trade Union Activism*, Sydney: Evatt Foundation.

Ewing, Keith D. 1993, 'Trade Union Derecognition and Personal Contracts: A Note on Recent Developments', *Industrial Law Journal* 22(4) December: 297–303.

Ewing, Keith D. and Napier, Brian W. 1986, 'The Wapping Dispute and Labour Law', *Cambridge Law Journal* 45(2) July:285–304.

Facts on File 1993a, 'Murdoch Repurchases "New York Post"', April 8:249.

—— 1993b, 'Murdoch Unveils Expansion Plans', September 9:671.

—— 1993c, 'Murdoch "New York Post" Repurchase Official', December 31: 978.

Fagan, Robert H. and Webber, Michael 1994, *Global Restructuring: The Australian Experience*, Melbourne: Oxford University Press.

Ferguson, Thomas and Rogers, Joel 1986, *Right Turn: The Decline of the Democrats and the Future of American Politics*, New York: Hill and Wang.

Ferner, Anthony and Hyman, Richard 1992, 'Introduction: Industrial Relations in the New Europe', in *Industrial Relations in the New Europe*, eds Anthony Ferner and Richard Hyman, Oxford:Blackwell.

—— (eds) 1992, *Industrial Relations in the New Europe*, Oxford:Blackwell.

Frances, Raelene 1993, *The Politics of Work: Gender and Labour in Victoria 1880–1939*, New York: Cambridge University Press.

Franklin, Bob 1997, *Newszak and News Media*, London: Arnold.

Freeman, Chris, Soete, Luc and Efendioglu, Umit 1995, 'Diffusion and Employment Effects of Information and Communications Technology', *International Labor Review* 134(4–5):587–603.

Frenkel, Stephen J. 1988, 'Australian Employers in the Shadow of the Labor Accords', *Industrial Relations Journal* 27(2) Spring:166–79.

—— 1990, 'The Incidence and Control of Technical Change in the Australian Printing Industry', *Economic and Industrial Democracy* 11: 39–64.

Frenkel, Stephen J. and Weakliem, David L. 1989, 'Worker Participation in Management in the Printing Industry', *Journal of Industrial Relations* 31(4) December:478–99.

Garneau, George 1993, 'He's Back', *Editor and Publisher* 126(14) April 3:9–10.

Gennard, John 1990, *A History of the National Graphical Association*, London: Unwin Hyman.

Gennard, John and Dunn, Steve 1983, 'The Impact of New Technology on the Structure and Organisation of Craft Unions in the Printing Industry', *British Journal of Industrial Relations* 21(1):17–32.

Gifford, Courtney D. 1986–87, 1988–89, 1996, *Directory of U.S. Labor Organizations*, Washington DC: Bureau of National Affairs, Inc.

Gilbert, Nigel, Burrows, Roger and Pollert, Anna 1992, *Fordism and Flexibility: Divisions and Change*, Basingstoke, UK: Macmillan.

Glaberson, William 1994, 'Press Notes: The *Daily Mail* Moves Up the Pricing Ladder, But is the *Post* Ready to Come Down?', *New York Times* July 4:39.

Glyn, Andrew 1992, 'Exchange Controls and Policy Autonomy: The Case of Australia, 1983–1988', in *Financial Openness and National Autonomy. Opportunities and Constraints*, eds Tariq Banuri and Juliet B. Schor, Oxford: Clarendon.

Glyn, Andrew, Hughes, Alan, Lipietz, Alain and Singh, Ajit 1990, 'The Rise and Fall of the Golden Age', in *The Golden Age of Capitalism. Reinterpreting the Postwar Experience*, eds Stephen A. Marglin and Juliet B. Schor, Oxford: Clarendon Press.

Golden, Miriam and Pontusson, Jonas (eds) 1992, *Bargaining for Change. Union Politics in North America and Europe*, Ithaca, New York: Cornell University Press.

Goldfield, Michael 1987, *The Decline of Organized Labor in the United States*, Chicago: The University of Chicago Press.

Goltz, Gene 1989, 'Special Report: The Workforce Reorganisation', *Presstime* 11(9) September:18–23.

GPMU (Graphical, Paper and Media Union) 1998, World Wide Web Home Page, www.gpmu.org.uk/hub.html.

Graham, Andrew 1997, 'The UK 1979–95: Myths and Realities of Conservative Capitalism', in *Political Economy of Modern Capitalism: Mapping Convergence and Divergence*, eds Colin Crouch and Wolfgang Streeck, London: Sage.

Gramsci, Antonio 1971, *Selections from the Prison Notebooks*, trans. eds Quintin Hoare and Geoffrey Nowell Smith, London: Lawrence and Wishart.

Graphix 1981a, 'PATEFA Membership is Vital', 78 January/February:7.

—— 1981b, 'Daily Goes On-Line', 83 July:8–9.

—— 1983a, 'PATEFA Calls for Help from Hawke', 102 April:1, 3.

—— 1983b, 'End of an Era for Whitlock and PANPA', 103 May:2.

—— 1984a, 'Verdict on New-Tech Row "Soon"', 112 March:1.

—— 1984b, 'Study into Technology', 106 August:3.

—— 1988a, 'PKIU Accused of Losing Direction', 162 July:3.

—— 1988b, 'News Plans $126m Plant for Advertiser', 165 October:5.

—— 1990a, 'Adelaide Boost', 180 Febraury:11.

—— 1990b, 'Murdoch $12m "Robot" order', 182 April:3.

—— 1992, '*You'll Be Stronger With PATEFA On Your Team*: PATEFA advertisement', 212 December:8–9.

—— 1993a, 'More Cybergraphic Ad Systems at News', 216 May:3.

—— 1993b, 'Industry's Bodies Revamp for Future', 217 June:1, 3.

—— 1993c, '*You Can Save Money With Us*: PATEFA Advertisement', 217 June:26–7.

—— 1994, 'Cybergraphic Transforms News Editorial at Chullora', 230 August:7.

Griffin, Tony 1984, 'Technological Change and Craft Control in the Newspaper Industry: an International Comparison', *Cambridge Journal of Economics* 8:41–61.

Grint, Keith 1998, *The Sociology of Work: An Introduction*, 2nd edn, Cambridge, UK: Polity.

Guild Reporter 1978a, 'New York Guild Strikes The *Post* After Company Breaks Off Talks', 45(14) August 25:1, 7.

—— 1978b, 'Strike Talks Stalled in New York', 45(15) September 8:1, 6.

—— 1978c, 'N.Y. Talks go to Washington?', 45(16) September 22:1–2.

—— 1978d, 'New York Strike is Ended; 1400 Out in Vancouver', 45(19) November 10:1, 7.

—— 1982a, '*Courier–Express* Closure Toll; Jobs of 370 in Guild', 49(16) September 10:1–2.

—— 1982b, '*Courier–Express* Shut by Cowles on Day Promised', 49(17) September 24:1, 6.

—— 1982c, 'Buffalo Focusing on Easing Impact of Paper's Closure', 49(18) October 8:3.

—— 1982d, 'Murdoch's Boston Plan: Cut Back Ad Staff', 49(21) November 24:1, 2.

—— 1982e, 'Murdoch Gets Concessions, *Herald–American*', 49(22) December 20:1, 4.

—— 1994a, 'Publisher Buys *Boston Herald* from Murdoch', 61(2) February 18:5.

—— 1994b, 'Commission Finds Labor Relations in "Dismal" State', 61(6) June 24:1, 4.

—— 1994c, 'Technology Demand Scraps Boston Agreement', 61(6) June 24:3.

—— 1994d, 'Guild Advances in Computer Age', 61(11) November 18:8.

—— 1994e, 'CWA, GCIU have Evolved with Advancing Technology', 61(9) September 23:1, 6.

—— 1994f, 'In Brief: Members Ratify at Boston Herald', November 18:2.

—— 1994g, 'San Francisco Strike Ends with Five-Year Pacts', 61(11) November 18:3.

Hammond, Eric 1988, 'TUC Crisis. Danger: Flying Sparks: Charlie Leadbeater Interviews Eric Hammond, EETPU leader', *Marxism Today* 32(9) September:12–15.

—— n.d., 'Introduction', in *Training For a Secure Future. Towards 2000*, eds EETPU/EESA, EETPU/EESA.

Hammond, Ian 1985, 'Ultra Modern Press Facilities for "'Tiser'"', *Graphix* 129 September:1.

Harvey, David 1990, *The Condition of Postmodernity: An Enquiry into the Origins of Cultural Change*, Oxford: Blackwell.

Hayden, Edward P. 1980, 'The Luddites Were Right', *Guild Reporter* 47(22) December 12:8.

Head, Brian (ed.) 1983, *State and Economy in Australia*, Melbourne: Oxford University Press.

Hogben, Sean 1994, 'No Yellow Brick Road at News', *Alliance* Autumn:17.

Hollingsworth, J. Rogers and Boyer, Robert (eds) 1997, *Contemporary Capitalism: The Embeddedness of Institutions*, Cambridge, UK: Cambridge University Press.

Howarth, Bob 1997, 'Down Under Metros Complete Huge Cyber Pagination Project', *Editor and Publisher* June 21:http://www.cybergraphic.com.au/news/index.htm.

Hyman, Richard 1988, 'Flexible Specialisation: Miracle or Myth?', in *New Technology and Industrial Relations*, eds Richard Hyman and Wolfgang Streeck, London: Basil Blackwell.

—— 1989, *The Political Economy of Industrial Relations. Theory and Practice in a Cold Climate*, Basingstoke: Macmillan.

Immergut, Ellen M. 1998, 'The Theoretical Core of the New Institutionalism', *Politics and Society* 26(1) March:5–34.

Insider 1991a, 'A Growing Family', 1 April/May:3.

—— 1991b, 'Marvels of Mile End', 1 April/May:6–7.

—— 1991c, 'Men of Mile End', 2 June/July:2.

—— 1991d, '80 Veterans Bow Out', 4 October/November:2.

—— 1991e, 'Mile End Move Brings Best Out Of Staff', 4 October/November:2.

—— 1991f, 'Reinventing the Wheel. It's TQM – or Communication by Another Name', 4 October/November:9.

—— 1992a, 'Compositor Mal Thrives On Challenge', 7 April/May:2.

—— 1992b, 'Changes at The Top – With Greater Flexibility the Aim', 9 August/October:1.

—— 1993a, 'End of ENS for Editorial', 11 January–March:8.

—— 1993b, 'Classies Swing into Cyber', 14 August/September:3.

—— 1993c, 'Comps Gear Up to Meet Ad Design Challenge', 14 August/September:14.

—— 1993d, 'For ENS, read END', 14 August/September:6–7.

—— 1994, 'Training Crucial to Company's Future', 19 October–December:8.

—— 1995a, 'Printing Contract Loss Costs 70 'Tiser Jobs', 21 April–June:2.

—— 1995b, 'New Era in the Composing Room', 22 October–December:2.

—— 1996, 'Language of Pagination', 21 April–June:2.

Jenkins, Simon 1979, *Newspapers: The Power and the Money*, London: Faber and Faber.

Jessop, Bob 1991, 'Thatcherism and Flexibility: The White Heat of a Post-Fordist Revolution', in *The Politics of Flexibility: Restructuring State and Industry in Britain, Germany and Scandinavia*, eds Bob Jessop, Hans Kastendiek, Klaus Nielsen and Ove K. Pedersen, Aldershot, England: Edward Elgar.

Jessop, Bob, Bonnett, Kevin, Bromley, Simon and Ling, Tom 1987, 'Popular Capitalism, Flexible Accumulation and Left Strategy', *New Left Review* 165 September/October:104–22.

Johnson, Carol 1989, *The Labor Legacy. Curtin, Chifley, Whitlam, and Hawke*, Sydney: Allen & Unwin.

Joint Standing Committee for National Newspapers (JSC) 1976, *Programme For Action*, Manchester: Withy Grove Press.

Jones, Tim 1987, 'Unconditional Withdrawal by Print Unions. Wapping Dispute Ends with NGA Decision', *The Times* 7 February:1–2.

Journalist 1986, 'AJA/PKIU Joint Statement', March:3.

—— 1988, 'Executive, PKIU Agree on New Technology', March:4–5.

Kaiser, Robert G. 1985, 'The Strike at The *Washington Post*', in *SOGAT Study Group Report on New Technology in the Newspaper Industry Following Visit to USA and Canada in May 1985*, Hadleigh, Essex: SOGAT '82.

Kalleberg, Arne L., Wallace, Michael, Loscocco, Karyn A., Leicht, Kevin T. and Ehm, Hans-Helmut 1987, 'The Eclipse of Craft. The Changing Face of Labor in the Newspaper Industry', in *Workers, Managers, and Technological Change. Emerging Patterns of Labor Relations*, ed. Daniel B. Cornfield, New York: Plenum.

Katz, Harry C. and Kochan, Thomas A. 1992, *An Introduction to Collective Bargaining and Industrial Relations*, New York: McGraw-Hill.

Kelly, James 1986, 'Revolution on Fleet Street. Eddy Shah and Rupert Murdoch Transform the Face of the British Press', *Time* March 17:50–6.

Kerr, Clark and Staudohar, Paul D. (eds) 1994, *Labor Economics and Industrial Relations: Markets and Institutions*, Cambridge, MA: Harvard University Press.

Kiernan, Thomas 1986, *Citizen Murdoch*, New York: Dodd, Mead.

Kinnock, Neil 1986, 'Law v. Justice at Stalag Wapping', *SOGAT Journal* April:8–9.

Klima, Rob 1997, 'Milestone Marks End of Project', *The Insider* August–September:2.

Knights, David and Willmott, Hugh (eds) 1990, *Labour Process Theory*, Basingstoke: Macmillan.

Knoke, David, Pappi, Franz Urban, Broadbent, Jeffrey and Tsujinaka, Yutaka 1996, *Comparing Policy Networks: Labor Politics in the U.S., Germany, and Japan*, Cambridge, UK: Cambridge University Press.

Kochan, Thomas A., Katz, Harry C. and McKersie, Robert B. 1986, *The Transformation of American Industrial Relations*, New York: Basic Books.

—— 1994, *The Transformation of American Industrial Relations*, Ithaca, New York: ILR Press.

Kohn, Peter 1994, 'News Leaves Room to Grow', *Graphix* 230 April:12.

Kopeck, Thomas W. 1978a, 'Secretary-Treasurer's Viewpoint: How Many Times Must a Man Look Up Before He Can See the Sky?', *Typographical Journal* 172(4) April:3.

—— 1978b, 'Secretary-Treasurer's Viewpoint: You Take my Life When you Take the Means Whereby I Live', *Typographical Journal* 173(1) July:3.

Koschnick, Wolfgang J. 1989, 'As I See It: An Interview with Rupert Murdoch', *Forbes* November 27:98–104.
Kurtz, Howard 1994, *Media Circus: The Trouble with America's Newspapers*, New York: Random House.
Labour Research 1993, 'Lowest Paid Face Wages Free For All', 82(1) January:23–4.
Lamb, Larry 1982, 'Sir Larry, On Being an Editor', *Graphix* 97, October:2, 4.
Lansbury, Russell D. and Bamber, Greg G. 1989, 'Technological Change, Industrial Relations, and Human Resource Management', in *New Technology. International Perspectives on Human Resources and Industrial Relations*, eds Greg J. Bamber and Russell D. Lansbury, Sydney: Allen and Unwin.
Lansbury, Russell D. and Davis, Edward M. 1992, 'Employee Participation: Some Australian Cases', *International Labour Review* 131(2):231–48.
Lash, Scott and Urry, John 1987, *The End of Organized Capitalism*, Wisconsin: University of Wisconsin Press.
—— 1994, *Economies of Signs and Space*, London: Sage.
Lawrence, John 1983a, 'Split Between AJA and PKIU was a Tragedy', *Journalist* June:5.
Leandros, Nicos and Simmons, Colin 1992, 'New Technology and Changing Industrial Relations in Greece: The Case of the National Newspaper Industry 1979–85', *Cyprus Journal of Economics* 5(1) June:25–44.
Leapman, Michael 1983, *Barefaced Cheek: The Apotheosis of Rupert Murdoch*, London: Hodder and Stoughton.
Littleton, Suellen M. 1992, *The Wapping Dispute: An Examination of the Conflict and its Impact on the National Newspaper Industry*, Aldershot: Avebury.
Lloyd, Clem 1985, *Profession Journalist: A History of the Australian Journalists' Association*, Sydney: Hale and Iremonger.
Locke, Richard M. and Thelen, Kathleen 1995, 'Apples and Oranges Revisited: Contextualized Comparisons and the Study of Comparative Labor Politics', *Politics and Society* 23(3) September: 337–67.
MacIntosh, Malcolm 1984, 'Technological Change at Advertiser Newspapers Limited', *Work and People* 10(3):29–37.
Mail and Guardian (South Africa) 1995, 'The Man Who Would Buy the World', August: 4–10:22.
Maraghy, Gerry 1989, 'Living with Murdoch! The *Boston Herald* Experience', *The Word* May:5.
Marglin, Stephen A. 1990, 'Lessons of the Golden Age: An Overview', in *The Golden Age of Capitalism. Reinterpreting the Postwar Experience*, eds Stephen A. Marglin and Juliet B. Schor, Oxford: Clarendon.
Marglin, Stephen A. and Schor, Juliet B. (eds) 1990, *The Golden Age of Capitalism: Reinterpreting the Postwar Experience*, Oxford: Clarendon.
Marjoribanks, Timothy forthcoming, 'The "Anti-Wapping"? Technological Innovation and Workplace Reorganisation at the *Financial Times*', *Media, Culture and Society*.
Markey, Ray 1984, 'A Comparison of Trade Union Responses to Technological Change in Britain and Australia', *Work and People* 10(2):29–37.
Marshall, Ray 1994, 'Organizations and Learning Systems for a High-Wage Economy', in *Labor Economics and Industrial Relations: Markets and Institutions*, eds Clark Kerr and Paul D. Staudohar, Cambridge, MA: Harvard University Press.
Martin, Roderick 1981, *New Technology and Industrial Relations in Fleet Street*, Oxford: Oxford University Press.

Mathews, John 1989, *Tools of Change: New Technology and the Democratisation of Work*, Sydney: Pluto.

—— 1994, *Catching the Wave: Workplace Reform in Australia*, St Leonards, NSW: Allen & Unwin.

McCammon, Holly J. 1990, 'Legal Limits on Labor Militancy: U.S. Labor Law and the Right to Strike since the New Deal', *Social Problems* 37(2) May:206–29.

McEachern, Doug 1990, *The Expanding State: Class and Economy in Europe Since 1945*, Hemel Hempstead, England: Harvester Wheatsheaf.

—— 1991, *Business Mates: The Power and Politics of the Hawke Era*, Sydney: Prentice Hall.

—— 1992a, 'Political Parties of Business: Liberal and National', in *Business–Government Relations in Australia*, eds Stephen Bell and John Wanna, Sydney: Harcourt Brace Jovanovich.

—— 1992b, 'Business Responses to Labor Governments', in *Business–Government Relations in Australia*, eds Stephen Bell and John Wanna, Sydney: Harcourt Brace Jovanovich.

McIlroy, John 1991, *The Permanent Revolution? Conservative Law and the Trade Unions*, Nottingham, England: Spokesman.

McLoughlin, Ian and Clark, Jon 1994, *Technological Change at Work*, 2nd edn, Buckingham: Open University Press.

McLoughlin, Ian and Gourlay, Stephen 1994, *Enterprise Without Unions. Industrial Relations In The Non-Union Firm*, Buckingham: Open University Press.

McMichen, Robert S. 1986, 'Letter to All ITU Officers and Members, re Merger of ITU and CWA', *Typographical Journal* 189(2) August:17.

MEAA (Media, Entertainment and Arts Alliance) 1998, World Wide Web Home Page: www.alliance.aust.com.

Melvern, Linda 1986, *The End of the Street*, London: Methuen.

Merrill, John C. 1995, *Global Journalism. Survey of International Communications*, 3rd edn, USA: Longman USA.

Metal Worker 1994, 'PKIU Members in Huge Vote for Merger', 15(8) November–December:1–2.

Metcalf, David 1990, 'Movement in Motion', *Marxism Today* September:32–5.

Milkman, Ruth 1997, *Farewell to the Factory: Auto Workers in the Late Twentieth Century*, Berkeley: University of California Press.

Millward, Neil, Stevens, Mark, Smart, David and Hawes, W.R. 1992, *Workplace Industrial Relations in Transition: The ED/ESRC/PSI/ACAS Surveys*, Aldershot, England: Dartmouth.

Mitchell, Daniel J.B. 1994, 'A Decade of Concession Bargaining', in *Labor Economics and Industrial Relations: Markets and Institutions*, eds Clark Kerr and Paul D. Staudohar, Cambridge, MA: Harvard University Press.

Mitchell, Richard 1997, 'Quantum Leap To New Era', *Insider* August–September:2.

Mitchell, Richard, Hart, Jonathan and Owen-Brown, Michael 1996, 'Drivers of Change', *Insider* 21 April–June:1–2.

Mitchell, Richard and Rimmer, Malcolm 1990, 'Labour Law, Deregulation, and Flexibility in Australian Industrial Relations', *Comparative Labor Law Journal* 12(1) Fall:1–34.

Moghdam, Dineh 1978, *Computers in Newspaper Publishing. User-Oriented Systems*, New York: Marcel Dekker, Inc.

Monk, Scott 1996, 'New Training Scheme for Newspaper People', *Insider* 23 October–November:5.

Moody, Kim 1988, *An Injury To All. The Decline of American Unionism*, London: Verso.

Morris, Gillian S. and Archer, Timothy J. 1992, *Trade Unions, Employers and the Law*, Oxford: Blackwell Law.

—— 1993, *Trade Unions, Employers and the Law*, 2nd edn, Oxford: Blackwell Law.

Mortimer, T. 1985, 'Advertiser Plans New Print Centre', *Printing Trades Journal* October/November:145.

Murdoch, Rupert 1986a, 'Rupert Murdoch Speaks on the War at Wapping: Interview by Mark Day', *The PANPA Bulletin* March:13–16.

—— 1986b, 'Why I Love America', *Graphix* 137 June:4.

—— 1988, 'Murdoch to Managers: Be Tough. An Interview with Rupert Murdoch', *U.S. News and World Report* March 7:56.

—— 1989a, '"The Technology of Freedom" Edited Text of the Walter Wriston Lecture, Manhattan Institute, New York, Thursday November 9, 1989', *The Weekend Australian* November 11–12:25–6.

—— 1989b, '"As I See It: An interview with Rupert Murdoch" by Wolfgang J. Koschnick', *Forbes* November 27:98–104.

—— 1990, 'Technology Will Serve Individuals More: Interview by Constance Gustke', *Fortune* March 26:120.

—— 1993, 1996, 1997, 1998, 'Chief Executive's Report', *News Corporation Limited Annual Report*, New York City: Taylor and Ives.

Myers Report (Committee of Inquiry into Technological Change in Australia) 1980, *Technological Change in Australia*, 4 vols, Canberra: Australian Government Publishing Service.

National Graphical Association 1984, *The Way Forward – New Technology in the Provincial Newspaper Industry*, Biennial Delegate Conference 1982: NGA.

—— 1986a, *News International Dispute*, Bedford: NGA (1982) Briefing Paper.

—— 1986b, 'Five new direct entry deals', *Print* 23(6) June:1.

National Union of Journalists (NUJ) 1977, *Journalists and New Technology*, London: Twentieth Century Press.

—— 1987, *National Union of Journalists Annual Report 1986–87*, London: NUJ.

—— 1998, World Wide Web Home Page, www.gn.apc.org/media/nuj.html.

Needham, Marian 1989, 'Dying with Murdoch! The *Buffalo Courier* Disaster', *Word* May:6–7.

Neil, Andrew 1996, 'Murdoch and Me', *Vanity Fair* 436 December:180–206

Nester, William R. 1998, *A Short History of American Industrial Policies*, New York: St Martin's Press.

News Corporation Limited 1997, 1998, *Annual Report*, New York City: Taylor and Ives.

—— 1997, 1998, *Financial Report*, Adelaide, Australia: The Griffin Press.

Newsom, Clark 1980, 'Newspapers, Labour Enter "New Era"', *Presstime* 2(8) August:4–9.

—— 1981, '"Buyouts" of Printers may have Peaked', *Presstime* 3(11) November: 42–3.

New Statesman and Society 1995, *Guide to Trade Unions and the Labour Movement 1995*, special supplement, London: Statesman and Nation Publishers.

Nieman Reports 1996, 'Cutbacks Squeeze Newspapers', 50(1) Spring:4–37.

Noon, Mike 1991, 'Strategy and Circumstance: The Success of the NUJ's New Technology Policy', *British Journal of Industrial Relations* 29(2) June: 259–76.

—— 1993, 'Control, Technology and the Management Offensive in Newspapers', *New Technology, Work and Employment* 8(2) September:102–10.

Oram, Steve 1987, 'Fleet Street Moves on: a Managerial Perspective', *Industrial Relations Journal* 8(2) Summer:84–9.

Organisation for Economic Co-operation and Development (OECD) 1994, *The OECD Jobs Study. Facts, Analysis, Strategies*, Paris:OECD.

Osborne, Billy 1990, 'Wapping . . . Now it's Time to Move in', *SOGAT Journal* June:25.

Ozaki, Muneto 1992, 'Technological Change and Labour Relations: an International Overview', in *Technological Change and Labour Relations*, ed. Muneto Ozaki, Geneva: International Labour Office.

Patrickson, Margaret 1986, 'Adaptation by Employees to New Technology', *Journal of Occupational Psychology* 59:1–11.

Pelling, Henry 1987, *A History of British Trade Unionism*, 4th edn, Harmondsworth, England: Penguin.

Pi 1967a, 'Computer Helping to Set Our Type', 125 January–February:1.

—— 1967b, 'Vast Press Room Changes Speed Production', 130 November–December:6.

—— 1974, '*The Advertiser* is First', 168 March–April:1.

—— 1977a, 'Introducing ENS', 186 March–April:1, 2, 15, 16.

—— 1977b, 'ENS Training Underway', 187 May–June:2.

—— 1977c, 'ENS Underway', 189 November–December:6–7.

—— 1979, 'ENS Training in Full Swing', 203 November–December:1.

—— 1981, 'Faith in System 5500', 212 May–June:6.

—— 1985a, 'Ex-Stereotypers qualify in second career', 230 June:5.

—— 1985b, 'Long-Term Plan will give *Advertiser* an Immense New Printing Centre', 231 September:1, 9.

—— 1989, 'Mile End Now a $133-Million Project', 244 June:12–13.

—— 1990, 'Meticulous Model Shows the Way Ahead', 247 June:5–6.

Pierson, Paul 1994, *Dismantling the Welfare State? Reagan, Thatcher, and the Politics of Retrenchment*, Cambridge, UK: Cambridge University Press.

Piore, Michael and Sabel, Charles 1984, *The Second Industrial Divide: Possibilities for Prosperity*, New York: Basic Books.

PKIU (Printing and Kindred Industries Union) 1987, 'Information Sharing', *PKIU Technology Guidebook*, Sydney: PKIU.

Police Monitoring and Research Group (PMRG) 1987, *Policing Wapping: An Account of the Dispute 1986/7. Briefing Paper Number 3*, London: Police Monitoring and Research Group.

Pollert, Anna 1988, 'Dismantling Flexibility', *Capital and Class* 34:42–75.

Poster, Mark 1995, *The Second Media Age*, Cambridge, MA: Polity Press.

Potter, Elizabeth J. 1988, 'UK Labour Law: A Retrospective Assessment of the "Wapping Dispute"', *International Business Lawyer* March:107–12.

Probert, Belinda 1993, 'Reconceptualising Restructuring: The Australian Workplace in a Changing World Economy', *Working Paper 1993/1* Melbourne: CIRCIT.

Printing Trades Journal (PTJ)1980, 'Adelaide and Hobart Agreements Certified', LXIV(1) January:5.

—— 1981, 'Hot Metal Ends at Advertiser', LXV(6) July:102.

—— 1987, ' "Newspaper Printing Agreement" – Negotiations Underway with Employers', LXXII(3) April/May:46–7.

—— 1990, 'News Corps Printing Industry Revolution', June Quarter:13–14.

—— 1992, 'Newspaper Unions Decide on Joint Approach', December Quarter: 12.

—— 1993a, 'News Ltd 10 Per Cent Deal Imminent', Winter/July:6.

—— 1993b, 'PKIU Steps in Over Threat of Sackings', Spring Quarter/October:5.

Purcell, John 1993, 'The End of Institutional Industrial Relations', *Political Quarterly* 64(1) January–March:6–23.

Raskin, A.H. 1978, 'It Isn't Labor's Day', *Nation* 227(7) September 9:197–201.

—— 1979a, 'A Reporter At Large: The Negotiation I: Changes in the Balance of Power', *New Yorker* January 22:41–87.

—— 1979b, 'A Reporter At Large: The Negotiation II: Intrigue at the Summit', *New Yorker* January 29:56–85.

Reed, Rosslyn 1988, 'From Hot Metal to Cold Type Printing Technology', in *Technology and the Labour Process: Australasian Case Studies*, ed. Evan Willis, Sydney: Allen and Unwin.

—— 1991, 'Professionalism and Information Technology: Age Journalists', in *Information Technology in Australia. Transforming Organisational Structure and Culture*, ed. Stan Aungeles, Kensington, New South Wales: NSW University Press.

Richards, G.W. 1986, 'Murdoch Deceit Unmasked', *London SOGAT Post* February:3.

Richardson, M.J. 1988, 'Don't Forget the Major Wapping Issues: Letter', *SOGAT Journal* February:10.

Rogers, Joel 1990, 'Divide and Conquer: Further "Reflections on the Distinctive Character of American Labor Laws"', *Wisconsin Law Review* 1990(1):1–147.

—— 1992, 'In the Shadow of the Law: Institutional Aspects of Postwar U.S. Union Decline', in *Labor Law in America. Historical and Critical Essays*, eds Christopher L. Tomlins and Andrew J. King, Baltimore: The Johns Hopkins University Press.

Ruffini, Gene 1989, 'Surviving Murdoch! The *New York Post* Sale', *The Word* May: 7–8.

Sabel, Charles 1982, *Work and Politics: The Division of Labor in Industry*, Cambridge, UK: Cambridge University Press.

Sabel, Charles F. and Zeitlin, Jonathan 1997, *World of Possibilities: Flexibility and Mass Production in Western Industrialisation*, Cambridge, UK: Cambridge University Press.

Sacked *Sunday Times* NUJ Reporters 1986, 'Insight Investigation. Wapping and the EETPU: The Faulty Connections', *Wapping Post* August 30:1–3.

Sahu, Anandi and Tracy, Ronald L. (eds) 1991, *The Economic Legacy of the Reagan Years: Euphoria or Chaos?*, New York: Praeger.

Scales, John 1991, 'Mile End's Problems Are Being Sorted Out', *Insider* 2 October/November:1.

Scargill, Arthur 1987, 'We are on the Edge of an Abyss', *Wapping Post* March 31:6.

Schor, Juliet and You, Jong-Il (eds) 1995, *Capital, the State, and Labour: A Global Perspective*, Brookfield, Vermont: Edward Elgar.

Scott, Daniel T. 1987, *Technology and Union Survival: A Study of the Printing Industry*, New York: Praeger.

Sharp, Margaret and Walker, William 1994, 'Thatcherism and Technical Advance – Reform Without Progress?' in *Britain's Economic Performance*, eds Tony Buxton, Paul Chapman and Paul Temple, London: Routledge.

Shawcross, William 1992a, *Murdoch*, New York: Simon and Schuster.

—— 1992b, *Ringmaster of the Information Circus*, London: Chatto and Windus.

Singleton, Gwynneth 1992, '"New Federalism" and Industrial Relations',

Australian Journal of Political Science, special issue 27:127–42.

Sisson, Keith 1975, *Industrial Relations in Fleet Street: A Study in Pay Structure*, Oxford: Blackwell.

Sklair, Leslie 1995, *Sociology of the Global System*, 2nd edn, Sydney: Prentice Hall.

Skocpol, Theda 1980, 'Political Responses to Capitalist Crisis: Neo-Marxist Theories of the State and the Case of the New Deal', *Politics and Society* 10(2):151–201.

Smith, Anthony 1980, *Goodbye Gutenberg. The Newspaper Revolution of the 1980s*, New York: Oxford University Press.

SOGAT 1985, *New Technology: The American Experience. Study Group Report on New Technology in the Newspaper Industry Following Visit to USA and Canada in May 1985*, Hadleigh, Essex: SOGAT '82.

—— 1986, 'No Stemming the Tide of Technology', *SOGAT Journal* January:11.

—— 1987, 'Leaked Documents Show Low Morale at Wapping', *SOGAT Journal* November:8–9.

—— 1990, 'Wapping . . . Now it's Time to Move in', *SOGAT Journal* June:25.

Steinmo, Sven, Thelen, Kathleen and Longstreth, Frank (eds) 1992, *Structuring Politics: Historical Institutionalism in Comparative Analysis*, Cambridge, UK: Cambridge University Press.

Stephens, Lowndes F. 1995, 'Media Systems: Overview', in *Global Journalism. Survey of International Communications*, 3rd edn, ed. John C. Merrill, New York: Longman USA.

Stewart, Andrew 1992, 'Procedural Flexibility, Enterprise Bargaining, and the Future of Arbitral Regulation', *Australian Journal of Labour Law* 5(2):101–34.

Sturdy, Andrew, Knights, David and Willmott, Hugh (eds) 1992, *Skill and Consent. Contemporary Studies in the Labour Process*, London: Routledge.

Thatcher, Margaret 1977, *Let Our Children Grow Tall: Selected Speeches 1975–1977*, London: Centre for Policy Studies.

—— 1986, 'Maggie's Brave New World: Letter to Robert Litherland MP', *Print* March:4.

The Times 1987, 'One Year On at Wapping', 27 January: Editorial.

Thomas, Robert J. 1994, *What Machines Can't Do: Politics and Technology in the Industrial Enterprise*, Berkeley: University of California Press.

Tolliday, Steven and Zeitlin, Jonathan 1991, 'Introduction: Employers and Industrial Relations Between Theory and History', in *The Power to Manage? Employers and Industrial Relations in Comparative and Historical Perspective*, eds Steven Tolliday and Jonathan Zeitlin, London: Routledge.

Tomlins, Christopher L. and King, Andrew J. (eds) 1992, *Labour Law in America. Historical and Critical Essays*, Baltimore: The Johns Hopkins University Press.

TUC (Trades Union Congress) 1979, *Employment and Technology*, London: TUC.

—— 1988, *The EETPU Suspension: A Trades Union Congress (TUC) Information Note*, London: TUC.

—— 1998, World Wide Web Home Page: www.tuc.org.uk.

Tucille, Jerome 1989, *Rupert Murdoch*, New York: Donald I. Fine.

Tunstall, Jeremy 1996, *Newspaper Power: The New National Press in Britain*, Oxford: Clarendon Press.

Upham, Martin (ed.) 1993, *Trade Unions of the World 1992–93*, 3rd edn, Harlow, Essex: Longman Current Affairs.

US Department of Labor Bureau of Labor Statistics 1994, *Occupational Projections and Training Data. Bulletin 2451*, Washington DC: US Government Printing Office.

Wapping Post 1987, 'We Will Never Forget', March 31:6.

Webster, Frank 1995, *Theories of the Information Society*, London: Routledge.

Wedderburn of Charlton, Lord 1989, 'Freedom of Association and Philosophies of Labour Law', *Industrial Law Journal* 18:1–38.

Weiler, Paul C. 1990, *Governing the Workplace: The Future of Labor and Employment Law*, Cambridge, MA: Harvard University Press.

Weinstein, Marc and Kochan, Thomas 1995, 'The Limits of Diffusion: Recent Developments in Industrial Relations and Human Resource Practices', in *Employment Relations in a Changing World Economy*, eds Richard Locke, Thomas Kochan and Michael Piore, Cambridge, MA: The MIT Press.

Wilby, Peter 1986, 'Behind Barbed Wire', *New Socialist* 36 March:31.

Williams, Claire with Bill Thorpe 1992, *Beyond Industrial Sociology: The Work of Men and Women*, Sydney: Allen & Unwin.

Willman, Paul 1986, *Technological Change, Collective Bargaining and Industrial Efficiency*, Oxford: Clarendon.

Wintour, Charles 1989, *The Rise and Fall of Fleet Street*, London: Hutchinson.

Wood, Stephen (ed.) 1989, *The Transformation of Work? Skill, Flexibility and the Labour Process*, London: Unwin Hyman.

Zeitlin, Jonathan 1985, 'Shopfloor Bargaining and the State: A Contradictory Relationship', in *Shopfloor Bargaining and the State. Historical and Comparative Perspectives*, eds Steven Tolliday and Jonathan Zeitlin, Cambridge: Cambridge University Press.

Zimbalist, Andrew 1979, 'Technology and the Labor Process in the Printing Industry', in *Case Studies on the Labor Process*, ed. Andrew Zimbalist, New York: Monthly Review Press.

Legal Cases

Cahill decision 1977. In *re John Fairfax and Sons Ltd re New Agreements* 1977 A.R. (Arbitration Reports) 905–11 (No. 344 of 1977).

Donaldson decision 1984. *In the matter of an application by the Printing and Kindred Industries Union to vary the Newspaper Printing Agreement 1981* CAR (Commonwealth Arbitration Reports) 463–8 (C No. 4010 of 1983) N007 Mis 343/84 MD Print F6660.

Termination, Change and Redundancy (TCR) Case 1984. In the matter of notifications of industrial disputes between *The Amalgamated Metals Foundry and Shipwrights' Union* and *Broken Hill Proprietary Co. Limited, Whyalla* and others (C No. 3690 of 1981), *Electrical Trades Union of Australia* and *Metal Trades Industry Association of Australia* and others (C No. 3735 of 1981), *Transport Workers' Union of Australia* and *Ansett Transport Industries (Operations) Pty Limited* and others (C No. 127 of 1983). 294 Commonwealth Arbitration Reports (CAR) 175–255 /Mis 250/84 MD Print F6230.

Index

DATE DUE

Demco, Inc. 38-293